Atomics in
the Classroom

ALSO BY MICHAEL SCHEIBACH

*In Case Atom Bombs Fall: An Anthology of
Governmental Explanations, Instructions and Warnings
from the 1940s to the 1960s* (edited by; McFarland, 2009)

*Atomic Narratives and American Youth:
Coming of Age with the Atom, 1945–1955* (McFarland, 2003)

Atomics in the Classroom

Teaching the Bomb in the Early Postwar Era

MICHAEL SCHEIBACH

McFarland & Company, Inc., Publishers
Jefferson, North Carolina

LIBRARY OF CONGRESS CATALOGUING-IN-PUBLICATION DATA

Names: Scheibach, Michael, 1949–
Title: Atomics in the classroom : teaching the bomb in the early postwar era / Michael Scheibach.
Description: Jefferson, North Carolina : McFarland & Company, Inc., Publishers, 2015. | Includes bibliographical references and index.
Identifiers: LCCN 2015039269| ISBN 9781476663562 (softcover : acid free paper) | ISBN 9781476622989 (ebook)
Subjects: LCSH: Atomic bomb—Study and teaching—United States. | Atomic bomb—Social aspects—United States. | Nuclear energy—Study and teaching—United States.
Classification: LCC QC773 .S34 2015 | DDC 355.02/170712—dc23
LC record available at http://lccn.loc.gov/2015039269

BRITISH LIBRARY CATALOGUING DATA ARE AVAILABLE

ISBN (print) 978-1-4766-6356-2
ISBN (ebook) 978-1-4766-2298-9

© 2015 Michael Scheibach. All rights reserved

No part of this book may be reproduced or transmitted in any form or by any means, electronic or mechanical, including photocopying or recording, or by any information storage and retrieval system, without permission in writing from the publisher.

Front cover: illustration of students reading bulletin board (Georgia Civil Defense Division)

Printed in the United States of America

McFarland & Company, Inc., Publishers
Box 611, Jefferson, North Carolina 28640
www.mcfarlandpub.com

To Archer, Scarlett, Joaquin, and Isabella,
America's next generation coming of age in a
new century filled with promise and opportunity

Table of Contents

Preface   1

Introduction   7

1. One World or None   15
   Teaching International Understanding   35
2. Teaching for the Atomic Age   45
   A Unit Outline for Teachers   67
3. Fear, Anxiety and Civil Defense   74
   Civil Defense in the Classroom   92
4. Safeguarding Democracy   115
   Educating for Survival   130
5. The New Frontier   143

Conclusion   151

Appendix A: Suggested Learning Experiences   159

Appendix B: Rural Civil Defense Youth Program   176

Appendix C: Glossary of Atomic Terms   183

Chapter Notes   193

Bibliography   203

Index   211

One continuing purpose of education is to provide children with the knowledge they will need to live successfully, productively, and happily. Man in the nuclear age now has the means at hand to destroy himself and his civilization. As more nations secure atomic weapons, the chance for irresponsible employment of these weapons is increased. While it may be devoutly hoped that nuclear warfare between the great nations of the world will never happen, it is the responsibility of our schools to educate our children for survival should such a disaster come.

James E. Allen, Jr.
Commissioner of Education, State of New York
Nuclear Survival: A Resource Handbook
(December 1960)

Preface

"The experiences of American kids in the Cold War were very different from those of their parents," historian Robert A. Jacobs has written. "While adults perceived a threat to the American way of life—to their health and well-being and those of their families—their children learned to fear the loss of a future they could grow into and inhabit. These kids of the Atomic Age wondered if they might be the last children on Earth."[1]

I am one of those atomic kids, born the same year the Soviet Union exploded its first atomic bomb, four years to the month after the United States dropped two atomic bombs—affectionately known as Fat Man and Little Boy—on Hiroshima and Nagasaki, Japan, in August 1945, helping to bring a world war still being fought on distant Pacific islands to a sudden and welcomed end. It was also the year China fell to Mao Zedong and the Communists; Alger Hiss went on trial as a Russian spy; and the United States and its allies formed the North Atlanta Treaty Organization (NATO), the first peacetime military alliance whose sole purpose was to stop Soviet aggression from overwhelming freedom-loving countries. One could argue that my coming into the world coincided with the official solidification of the Cold War, which had been slowly congealing since Communist Party General Secretary Joseph Stalin of the Soviet Union, British Prime Minister Winston Churchill, and U.S. President Harry S. Truman met at the Potsdam Conference in occupied Germany from July 27 to August 2, 1945—just days before the United States dropped the first atomic bomb—to finalize plans for dividing up the postwar world.

I entered kindergarten two years after the United States again flexed its nuclear muscles with the detonation of a hydrogen bomb in November 1952, and only a year after the Soviet Union matched this feat with a successful test of a hydrogen bomb in August 1953. The hydrogen bomb outpunched the first atomic bombs that killed some 200,000 Japanese men, women, and children by a thousand-fold. Compared to Hiroshima and

Nagasaki, where both people and buildings survived the blasts, the U.S. test conducted on the Pacific island of Elugelab in the Eniwetok atoll resulted in a cloud measuring one hundred miles wide and twenty-seven miles high. The bomb completely vaporized the island and created a hole in the atoll more than a mile wide and deep enough to hold the 103-story Empire State Building.[2]

I began my high school years shortly after President John F. Kennedy stared down Soviet Premier Nikita Khrushchev in October 1962 over the installation of intercontinental ballistic missiles in Cuba, only ninety miles from the U.S. border and reportedly armed with nuclear warheads that could reach American cities along the Eastern seaboard, including Washington, D.C. The thirteen-day confrontation became the first—and ultimate—reality television program. Addressing the nation on October 22, President Kennedy declared Cuba to be a Soviet strategic base and a threat to American security. He explained to a captive—and anxious—television audience that he was implementing a seven-part course of action, including the quarantine of all offensive military equipment being sent to Cuba and the demand that Khrushchev cease his current course of action. "Our goal is not the victory of might," Kennedy declared, "but the vindication of right—not peace at the expense of freedom, but both peace and freedom, here in this Hemisphere, and, we hope, around the world."[3]

As with most kids during this era, I sat through many class periods and attended many school assemblies learning the importance of protecting democracy and fighting communism. And, of course, I practiced "duck and cover" drills beginning in kindergarten. Ducking under my school desk never fazed me, at least in retrospect. My classmates were under their desks, too, so I assumed we'd all live or all die together. In high school—because we were too big to get under our desks, I suppose—we practiced shelter drills exclusively. During these drills, we walked single file down long corridors and descended into the dark recesses of the school, finally entering a cavernous room with large metal drums and boxes of survival food and water stacked floor to ceiling. These supplies were not just for us, though. Our high school also provided protection for the neighborhood, and brandished a bright yellow Fallout Shelter sign on each outside door to alert the surrounding neighborhood about where to go … just in case.

As an inner-city kid living in a walk-up apartment (i.e., no elevator), I also participated in spontaneous building drills. Without warning, an alarm sounded in the hallway to signal the necessity for my mother and me to stop whatever we were doing—immediately—and walk down three flights of stairs as quickly as possible to the protection of the cement-walled basement. There, all the tenants would gather around, make small talk, and eventually return to their favorite television programs, such as *Ozzie and Harriet* and *Gunsmoke*. Another atomic disaster averted. Until the next Operation Alert, that is. Instituted by the Federal Civil Defense Administration in 1954, Operation Alert was a fictitious atomic attack held annually until it ended in 1961. Each year, an all-out Soviet atomic attack killed millions of Americans while devastating the country's major metropolitan areas. Fortunately, it also afforded adults and children alike, including me, the opportunity to volunteer as pretend survivors, complete with burn wounds, broken limbs, and numerous ill effects from an atomic blast.

My childhood also witnessed Sputnik, the first artificial satellite launched in 1957 by the Soviet Union, which rocked America's confidence in its scientific mastery. After school, I watched the likes of Superman, Flash Gordon, Captain Video, and Tom Corbett, four television superheroes who battled evil villains on Earth and from outer space. My comic books transformed atomic warfare into highly entertaining graphic storylines, with titles such as *Atomic Spy*, *Atomic Attack*, *Atomic War*, *World War III*, and *Atomic-Age Combat*. And I spent many a Saturday afternoon in a darkened movie theater gripping my seat as an onslaught of gargantuan, atomic-mutated insects and alien beings from faraway planets wielding atomic blasters scared the wits out of me and the rest of the mesmerized, prepubescent audience. No one could resist buying a 25-cent ticket to be traumatized by movies such as *The Atomic Man*, *Forbidden Planet*, *Attack of the Crab Monsters*, *Not of This Earth*, *The Blob*, *The Fly*, and *War of the Worlds*. Even more frightening than mutant insects and space aliens, however, were the ever-so-deadly effects of radiation—especially in the 1957 movie *The Incredible Shrinking Man*. When Scott Carey (aka actor Grant Williams) begins to shrink after his cabin cruiser passes through a radioactive cloud off the California coast, then fights for his life as an inch-high man against a voracious spider, and ultimately vanishes into nothingness, it was enough to accentuate the atomic fears of everyone in the audience under the age of 10—and even older.

Needless to say, by my teenage years I had adopted, along with most of my friends, the attitude of the original atomic kid himself, *Mad* magazine's Alfred E. Neuman, whose motto was simply "What, me worry?" Launched as a comic book in 1952, *Mad* evolved into a magazine format in 1955 and officially introduced Alfred E. Neuman on its December 1956 cover—the perfect response to President Dwight D. Eisenhower's military strategy based on "mutually assured destruction." As a means of forgetting whatever worries I did have during my early teenage years, my buddy, who lived down the hall, and I would venture a few blocks from our apartment building after dinner, meander down a trash- and stench-filled alleyway, and climb up on the roof of the neighborhood bowling alley. There, outstretched and trouble-free, we'd watch Telstar, the first American communications satellite launched in 1962, drift lazily across the night sky, oblivious to all the fear and anxiety down below.

As Jacobs and numerous other historians point out, coming of age during the early Cold War era presented America's school-age children with myriad images and messages that helped shape their attitudes and outlooks about the present and the future. From drills to protect against atomic annihilation to Superman saving the world; from television's Wild West gunslingers on *The Rebel* and *Have Gun Will Travel* to Alan Shepherd and John Glenn hurtling through space; and from intercontinental ballistic missiles ready to launch nuclear warheads to the celestial fins and irradiant chrome on Detroit's ultra-cool automobiles, it was a complex era to say the least.

My interest in this period resulted in my first book, *Atomic Narratives and American Youth: Coming of Age with the Atom, 1945–1955*, which examines the impact of the atomic bomb on America's youth at home, at school, and in the community. This book then led to my second book, *"In Case Atom Bombs Fall": An Anthology of Governmental Explanations, Instructions and Warnings from the 1940s to the 1960s*, which documents the government's campaign to inform, to educate, and, arguably, to scare Americans about the real probability of a nuclear attack by the Soviet Union.

Atomics in the Classroom: Teaching the Bomb in the Early Postwar Era builds on the previous books but focuses on elementary and secondary education during the first two decades of the atomic age. Specifically, it explores how atomic issues (particularly "the bomb") became essential subjects in the curriculum of America's public schools beginning in the fall of 1945, shortly after the atomic bomb brought an abrupt conclusion to a protracted, deadly world conflict that had begun nearly a decade earlier, and continuing well into the 1960s, until more urgent concerns, primarily civil rights and Vietnam, redirected the nation's attention.

My objective is to provide not only a better understanding of how the government and educators viewed the necessity of preparing America's younger generation to confront the dangers and the potential of the new atomic age; it also is to demonstrate how civil defense drills, albeit integral to a school's regular activities, constituted only one aspect of a broader agenda that incorporated atomic topics into every classroom at every grade level practically every day.

As we continue our journey into the twenty-first century, with its own plethora of dangers and threats, it is imperative that we prepare today's children to cope in the post–9/11 world. From terrorist acts by foreign enemies, to acts of violence by domestic terrorists and fanatics, to the proliferation of nuclear weapons among antagonistic nation-states, America's elementary, middle, and high school students today face a far different world than those enrolled in schools from the late 1940s through the 1960s. Yet the fears and anxieties are the same. Perhaps closer insight into how America's educational system dealt with the threat of nuclear annihilation during a heated Cold War between two superpowers will provide a better perspective of our schools' challenges today.

Introduction

Only if our youth is made fully cognizant of its added responsibilities as citizens in the newly evolving atomic era can we be assured of the will of our people to resist aggression and the ability of our people to survive its disastrous effects.—
Clarence R. Huebner, Director, New York Civil Defense Commission[1]

In February 1963, just four months after the United States had teetered on the brink of nuclear war with the Soviet Union over missile installations in Cuba, Elsa Knight Thompson, director of public affairs for San Francisco radio station KPFA, interviewed four sixth-graders from San Bruno, California. The moderator's questions focused on the children's views of the atomic bomb and the "duck and cover" drills that had become embedded activities in elementary and secondary schools throughout the nation, from rural areas and small towns to the suburbs and inner cities.[2]

KPFA's interview took place twelve years after "duck and cover" drills began on a widespread basis in 1951, less than two years following the Soviet Union's successful test of an atomic bomb in August 1949 and the same year the Federal Civil Defense Administration (FCDA) began its mission to inform and to educate Americans of all ages—particularly school-age children and their parents—about how to prepare for and survive an atomic war. New York City was among the first cities to conduct "sneak attack drills," which were introduced February 7, 1951, in the city's 850 schools. By the end of 1952, close to ninety percent of America's schools from coast to coast practiced some form of civil defense training on a regular basis, with public, private, and parochial schools also distributing identification or so-called "dog tags" to children, complete with name, age, and blood type. As historian JoAnne Brown points out, civil defense in the 1950s became a way of life in American schools as educators embraced President Harry S. Truman's proclamation, "Education is our first line of defense."[3]

"Duck and cover" drills did not encompass the totality of schools' efforts to prepare children for what was considered the real probability of an atomic attack, however. Rather, these drills, although considered essential, were merely part of a wider commitment to educate America's students about the duplicity of the atomic age: the threat of the atomic bomb versus the peaceful potential of the atom. Even before the Soviet Union's acquisition of the atomic bomb, in fact, America's schools had begun to incorporate atomic themes into the curriculum. Aaron Goff, a junior high school teacher in Newark, New Jersey, has been credited with naming this new approach "atomics," which he defined as integrating atomic topics into all school classes—including history, science, home economics, mathematics, citizenship, English, and art—in order to prepare students to cope in the Atomic Age.[4] Commenting on atomics in 1950, educator Willem J. Van Der Grinten placed the responsibility for "coherent atomic education of our young people" squarely on America's teachers. According to Van Der Grinten, atomic energy issues needed to be addressed by teachers in all disciplines because "the consequences and implications of atomic energy … affect our civilization more than anything else."[5]

These consequences and implications were quite evident in the comments of the students interviewed by Thompson. When she asked Susan, one of the sixth-graders, if "duck and cover" drills were beneficial, she responded simply, "No." Then she went on to say what many students, as well as many educators and parents alike, had come to believe by the early 1960s.

"I don't really think these drills would do much good at all," she said, "because in such bombings as an atom bomb and if it was as close as one hundred miles off, the radiation still would in time reach you, and you couldn't live through that if you were under your desk and if you were on the roof or anything like that. I don't believe it would help at all—to be under your desk." Kathy, Susan's classmate, agreed, saying, "It won't help much to be in there under your desk or in a shelter because even if you survived radiation, it would be horrible to be like there is only a few people on earth."

When asked whether someone could live through an atomic bomb attack, Susan again expressed the feelings of many others nearly twenty years following the atomic bombings of Hiroshima and Nagasaki, Japan, in August 1945: "Even if a bomb was actually to be dropped on the United

States, I should probably not feel too badly about it because everybody would probably—many of the other people would be dying—it would only seem right; it wouldn't seem right to me at all if I were one of the only people who lived. And so, I would really prefer to die in a bombing than to live if most of the people were dying."

Kathy added that death was not frightening to think about unless you were in "horrible pain." Then she continued, "[I]f it comes to something like starvation, some do survive and something like—some must die and some mustn't … well, I'd rather die and let some person who would be more helpful to whoever survives—like a doctor—I'd rather die instead of having a doctor or learned person who knows about these things—I'd rather die instead of they."

Robert, one of two boys interviewed, agreed that he would rather die, too, if it meant a more essential person would live. "But as long as there is some hope for survival," he added, "I'd try to keep on." Fred, the fourth sixth-grader interviewed, also wanted to live: "Well, chances are you could do something to really help the world and if you die, you wouldn't do anything and couldn't help anybody."

These four students were not only aware of the atomic threat; they had obviously thought about it, understood the Soviet Union's capabilities and current world situation, and had the ability to express their opinions based on knowledge about the atomic bomb, something that might seem beyond the scope of a sixth-grade student today. Yet in a study published in 1962, psychologist Sibylle Escalona reported that as soon as their parents and the adults around them become concerned about such issues as atomic fallout, testing, and shelter building, children also learned about the atomic threat—as young as four years old. "Even small children," she wrote, "understand that nuclear weapons might really be used, on purpose and by people with intent to hurt. In this respect, their fearful thoughts about nuclear war are quite different from fears about monsters or about thunderstorms."[6] This understanding was nothing new, though. Often quoted, for example, is the child's prayer shortly after Hiroshima, "Please God, let us all perish in the same catastrophe."[7]

Escalona also suggested that knowing weapons capable of ending civilization exist yet often hearing that shelters and drills provide minimal protection, as well as observing the military and space successes of the Soviet Union, give children an image of the world as a dangerous and precarious place. Moreover, when children ask questions about the nuclear threat that go unanswered by adults, they feel helpless because they do not know what to expect, and, as a result, their sense of danger escalates into fear, which increases with heightened tension and anxiety among adults. Children also connect this precarious and frightening situation between the United States and the Soviet Union to the language of nuclear war (e.g., fallout, Russia, radiation, and H-bomb), which was evidenced by Susan, Kathy, Fred, and Robert. Escalona concluded, "We are the first generation to hold a veto power over continuing human life on earth; but we are also the first to be so fully capable of deciding our future."[8]

This comment, made in 1962, reflected the comments expressed by countless government officials, scientists, psychologists, journalists, and educators since World War II ended in 1945: namely, that atomic energy held both the promise of human betterment and the prospect of certain oblivion unless nations learned to live together in peace. One of the first commentaries on the impact of the atomic bomb was written by Norman Cousins, the thirty-three-year-old editor of the *Saturday Review of Literature*. In an editorial titled "Modern Man Is Obsolete," published August 18, 1945, less than two weeks after the bombing of Nagasaki, Cousins proclaimed that the new atomic age had brought more fear than hope. "It is a primitive fear," Cousins wrote, "the fear of the unknown, the fear of forces man can neither channel nor comprehend. This fear is not new; in its classical form it is the fear of irrational death."[9] When *The New Yorker* dedicated its entire August 31, 1946, issue to John Hersey's article "Hiroshima," Americans' fears translated into a compulsion to read the stories of six survivors of the atomic bomb. The issue not only sold out, but Hersey's story was rebroadcast on radio programs, was published as a book that fall, became an international bestseller, and has remained in print ever since.[10] As if this wasn't enough to absorb, A. M. Holladay, of the George Peabody College of Teachers, exacerbated whatever fears Americans had by warning that same year that "a Hitlerian maniac" armed with atomic weapons could end all of civilization, even if it meant destroying himself.[11]

The culmination of this hyperbolic rhetoric in both the mass media and in academic journals created what historian Paul Boyer has called the "primal fear of extinction," which spread quickly after the end of the war.[12] In his book *By the Bomb's Early Light*, Boyer points out that although most Americans had an immediate positive response to the atomic bombings of Hiroshima and Nagasaki because it forced Japan to surrender, Americans soon became consumed with fear of what would happen in another war fought with atomic weapons, especially the impact on their children. Hersey, who described child survivors and ended his story about a 10-year-old girl survivor, had written, "It would be impossible to say what horrors were embedded in the minds of the children who lived through the day of the bombing in Hiroshima."[13] These unimaginable horrors created comparable fears among Americans for their own children and, Boyer concludes, "contributed to the larger uneasiness that seeped through the culture in the weeks after August 6, 1945."[14] On a wider scale, historian Margaret Gowing has argued that the atomic bomb dropped on Hiroshima served as a demarcation line in history "so that the centuries before August 6, 1945, were sharply separated from the years to come."[15]

This fear of mass destruction presented an especially critical challenge to educators because of their role in preparing youth for the future—a future in which the only choice, the only hope, was to save the world from atomic destruction. David Lilienthal, chairman of the U.S. Atomic Energy Commission, made this point in an article published in *Senior Scholastic* magazine, distributed to secondary schools throughout the country. Lilienthal stressed that the world can, indeed, "commit wholesale suicide" if the atomic bomb gets out of control, or that it can embrace a "brave new world" of a more abundant life. The choice, he said, rested in "our hands," which everyone reading *Senior Scholastic* understood as being the hands of teachers and their students.[16]

Edgar Dale of Ohio State University, writing in the October 1946 issue of *The High School Journal*, argued that schools had failed to "prevent the erosion of the human spirit" and did not meet the needs and demands of the atomic age. The most important educational job now, he argued, "is to learn to live decently with the people in our block, our city, our state, our nation, and the rest of the world." The country needed not only an educational system that met the challenges of the atomic bomb, but also one that ensured the security "that comes when common people develop a common approach to a common danger."[17]

Also writing in 1946, educator and textbook author C.S. Kazdan called for new educational methods that would address the needs of what he called "a new man" who would need to turn away from chauvinism, Nazism, and fascism. "Only with education based upon and inspired by far-reaching ideas and solutions that find their source in free associations of parents and teachers," he wrote, "can education make for a regenerated mankind, for the emergence of a new generation that would erect and uphold a peaceful and law-abiding society."[18]

Kazdan called the new educational curricula "the science of nations," designed to instill a personal sense of world citizenship and brotherhood. The new curricula was necessary in order to support the United Nations Education, Scientific and Cultural Organization (UNESCO), created in November 1945 for the purpose of, as stated in its constitution, "advancing, through the educational and scientific and cultural relations of the peoples of the world, the objectives of international peace and of the common welfare of mankind." Kazdan contended that these objectives would only be achieved if a renewed educational movement promoted the teaching of One World because, he wrote, "the problems for the future are fundamentally the same for the vanquished as well as for the victorious nations." The slogan of the new school, he said, must be "Know your neighbors! Learn to know the peoples of the world!"[19]

Kazdan and Dale, among many educators in the late 1940s, believed that world peace could only be achieved through the emphasis by schools on international brotherhood and One World. Yet by the 1950s, with the Soviet Union's atomic threat now a reality, America's schools expanded their curricula to include civil defense with the support of the Federal Civil Defense Administration's Civil Defense Education Project, which, in cooperation with the U.S. Office of Education, directed the public school program under the slogan "Education for National Survival." The project had two fundamental aims: to communicate directly with students by providing materials such as syllabi, curricula, films, and safety checklists for shelter life; and to communicate indirectly with parents by using children as interlocutors. The FCDA had been organized by Executive Order 10186 on December 1, 1950, and established as an official government agency on January 12, 1951, when President Harry S. Truman signed the Federal Civil Defense Act of 1950 into law.[20] Over the next decade, it published some 476 million pieces of literature, with forty percent directed at what it termed "home kits," targeted at parents and their children and designed to educate them on the nuclear threat and to be better prepared to survive an atomic war. For example, *Interim Civil Defense Instructions for Schools and Colleges*, published in 1951, argued that civil defense included everyone, not just certain groups. The brochure went on to recommend that educators not overlook opportunities to provide students with homework that involves "parents in training."[21]

FCDA administrator Clara McMahon addressed the inclusion of civil defense within the atomics curriculum in a 1953 article published in *The Elementary School Journal*. She argued that any threat to national security made it imperative for schools to adjust their curriculum to instill in their students the qualities and characteristics needed to persevere in such an emergency. She wrote, "When these goals are compared with the concepts of civil defense—individual self-protection, extended self-protection, mutual aid, and mobile support—it requires no leap of logic to see how closely the concepts can be tied in with the aims of education; indeed, that they are actually an extension of these aims." According to McMahon, teachers needed to emphasize the importance of controlling atomic energy to safeguard the world against atomic war and must make clear the harsh reality that no place, no matter how remote, is secure from attack. Most important, though, teachers must be sure their students understand the nation's very survival was at stake in what she called "the current world's struggle."[22] As historian William Graebner has argued, elementary and secondary schools became the locus in the nation's efforts to shape postwar youth in their behaviors and values, including their knowledge of and attitudes toward atomic energy.[23] Another historian has suggested that children constituted a new class of soldiers: "deterrence soldiers beckoned to peacetime behavior that was in concert with war prevention."[24]

Edgar Dale had articulated the critical mission of education in the atomic age in 1946, writing, "There is no choice. We either improve the quality of teaching and relate our schools to life—or else. The 'or else' is pretty clear now."[25] Fourteen years later, in 1960, James E. Allen, Jr., commissioner of education for the State of New York, echoed the same sentiments, writing, "While it may be devoutly hoped that nuclear warfare between the great nations of the world will never happen, it is the responsibility of our schools to educate our children for survival should such a disaster come." Allen argued that if teachers presented the facts about nuclear bombs as well as dangers associated with nuclear fallout in what he called "matter-of-fact terms," children would "more nearly assimilate in calm and deliberate ways the instruction deemed essential both for our country's security and for their own lives."[26]

Many historians of education during this period, however, have tended to direct their studies toward pedagogical issues, such as the postwar shift to "life adjustment" education, or "duck and cover" drills and related civil defense activities, while overlooking the emphasis by educators on the critical importance of atomic education, the incorporation of atomic themes throughout the school day—in classroom after classroom, and the impact of atomics on America's elementary and secondary students.

Beginning with an emphasis on One World or None and the international control of atomic energy in the late 1940s, educators didn't hesitate to fulfill their responsibility to prepare school-age children for the atomic age. Chapter 1 chronicles these early postwar years when scientists, educators, and government officials warned of the impending destruction not only of the United States but of Earth itself if another war—an atomic war—occurred. From Albert Einstein and members of the Manhattan Project, to David Lilienthal of the Atomic Energy Commission, to John W.

Studebaker, U.S. Commissioner of Education, America's teachers heard the same message: a call to arms to prepare their students for the promise and the perils of the atomic age. Chapter 2 moves into school halls and classrooms to look at the new curriculum incorporating atomic issues into all subject areas, including history, science, English, mathematics, home economics, and other classes. It also explores the extracurricular activities of children involving atomic energy, including the landmark traveling exhibit Alert America, which encouraged parents and educators to have children attend "The Show That Could Save Your Life," as it was billed.

Fear management and panic prevention became a central objective of the federal government in the early postwar years. Chapter 3 explores concerns about the impact of children's fears on their ability to cope in the atomic age. It also discusses how educators adopted the atomics curriculum in the belief that by understanding all aspects of atomic energy—its destructive and constructive applications—these fears would be lessened. Expanding beyond mitigating fear among children, Chapter 4 discusses the importance for school-age children to adhere to the tenets of democracy and adopt the traits of good citizenship. This emphasis began in the early postwar years and gained momentum in the 1950s, as the Soviet Union and the communist threat became much more pronounced. Chapter 5 then concludes this study by examining the early years of President John F. Kennedy's New Frontier, when the United States and the Soviet Union stood toe-to-toe in a nuclear standoff, as educators continued to follow the atomics curriculum for children from kindergarten through high school.

This study acknowledges the presence and importance of teen towns and sock hops, rock 'n' roll and drive-in movies, cashmere sweaters and ducktails, vanilla malts and cherry phosphates, as well as the reality of racial discrimination, juvenile delinquency, the rise of suburbia, and the great expanse of the 1950s cultural, social, and political landscape. These topics, however, have been well covered by others. Rather, this study focuses on the fact that nearly two generations of children—those enrolled in elementary and secondary schools at the close of World War II, and those entering school in the 1950s, such as the four sixth-grade students interviewed by San Francisco radio station KPFA in 1962—confronted the same prospect of nuclear annihilation, only escalated over time. And through atomics, incorporating atomic energy and its inimical repercussions into all school subjects, educators sought to help America's school-age children deal with their fears, become better citizens, and, if needed, survive a nuclear war.

1

One World or None

The terrific blast at Hiroshima shocked the world into a realization that catastrophe lies ahead if war is not eliminated. This great fear has for the time being overshadowed the hope that atomic energy may vastly enrich human life if given a chance.—Arthur Compton, Head, Metallurgical Laboratory, Manhattan Project[1]

During the years immediately following the end of World War II, educators presented their students with two distinct but related viewpoints: One World, replacing nation-states with a world government deemed essential to preventing a possible nuclear Armageddon, and, in contrast, individual nations maintaining their sovereignty but pledging their support for international cooperation through the United Nations. Only as the Soviet Union intensified its threat, particularly after 1949, did the focus shift more toward safeguarding democracy and defeating communism as the best means of ensuring world peace. Yet even then, the atomic apocalypse resided as a not-so-subtle subtext. Whether teaching One World or the safeguarding of democracy, however, teachers in the late 1940s never forgot their ultimate mission: to prepare American's children to confront and to survive in a brave new atomic world. This mission, in fact, formed the underlying basis of the paradigm shift in postwar education from progressive, or child-centered, education, as espoused by John Dewey, which had been dominant since its introduction during the Progressive Era at the turn of the twentieth century, to life-adjustment education, which shifted the focus from children being educated to help shape a democratic society, to children being educated to adjust to the demands and constructs of the existing society, which now found itself in a potentially catastrophic atomic age.[2] According to historian Andrew Hartman, postwar educators emphasized a form of therapeutic adjustment based on relevance, instrumentalism, social order, and patriotism. "[O]ne of the best ways in which the schools served as an instrument of national security was to inculcate patriotism in its young charges," he writes. "Although these objectives were latent to earlier progressive educational theory, they were made manifest during World War II and the Cold War."[3]

In the atomic age, patriotism itself became redefined to mean protecting not only the nation but also civilization, as demonstrated by the two emerging viewpoints of One World and international cooperation. Literally within days following V-J Day, journalists and social commentators opened their volleys: some arguing vehemently for the abolishment of nation-states, replacing them with a world government; others, realizing the utopian nature of One World, supporting the United Nations as the best hope for ensuring world peace. One editorial stated bluntly that the atom "belongs in the hands of an international police and nowhere else."[4]

The reasoning behind and justification for these two positions comprised the essence of *One World or None*, published in 1946 and promoted as a report to the public on the meaning of the atomic bomb. The book featured essays by the nation's preeminent atomic scientists, many of whom worked on the Manhattan Project to develop the atomic bomb. Albert Einstein, Harold Urey, J.R. Oppenheimer, Arthur Compton, Frederick Seitz, Hans Bethe, and Leo Szilard, among others, presented the only possible options to the prevention of an atomic-ladened World War III: either individual nation-states had to be abolished in favor of a world government, or the development of atomic energy had to be placed under international control. Urey, who had directed the Columbia University project during the war that developed a means of separating U-235 by using gaseous diffusion—U-235 being the key ingredient in the Little Boy atomic bomb dropped on Nagasaki—captured the overriding message from this landmark compilation. He wrote, "A world war in which atomic weapons are used might very well weaken all of our countries and peoples to such an extent that they would not be able to survive in the future. And not only may our own culture be destroyed by these weapons of mass destruction, but all civilizations as they exist in the world may be retarded and weakened for centuries to come. It all adds up to the most dangerous situation that humanity has ever faced in all history."[5] Americans responded enthusiastically to *One World or None* by buying some 100,000 copies and making it a *New York Times* bestseller.

Educators got the message, as well. Lewis Todd of the Danbury (Conn.) State Teachers College, for example, reflected the views of many teachers and administrators when he wrote in *School and Society*, "[T]he revolution which the atomic bomb has precipitated in the world of science must be matched by a revolution of equal force in the world of human relationships." The world had entered the final stretch of the "dreadful race between education and catastrophe" as the atomic bomb unleashed terror on the world. At the same time, the ability to control atomic energy had become high stakes in the game of life and death. He urged teachers to face the challenge of the new atomic world by underscoring to their students the importance of brotherhood, understanding, and friendship. It was no longer sufficient to enable youth to study about things, according to Todd, but to teach them to act intelligently.[6]

Without the broadening of educational objectives, wrote W. H. McFarland, supervisor of the Iowa Department of Public Instruction, there would be no world. The atomic bomb had issued a clarion call to all educators, he said. Schools must begin to teach world history, world geography, world economics, world social problems, world literature, and world cultures, all designed to foster international understanding and goodwill. Teaching the facts was not sufficient, however. It was also essential to unleash the emotions, "the heart," of youth in order to prepare them for the factual realities of the new age:

> We must teach "one world." But first we must feel "one world." Our "heart" education has not by any stretch of the imagination kept with up our "head" education. The effective approach to wholesome human relations must be an emotional approach. We must set the stage for the "one world" drama emotionally.... We need to cultivate the proper emotional tendencies—to train the heart. Then the heart should point the way, and the head should discover the means of getting there.[7]

Although McFarland admitted that training the heart savored propaganda, he argued it was "a good savor," an introductory step, a predisposition, to more intellectual education, which would stress the necessity of world government and world peace. "There must be 'one world,'" he wrote, "or there will be none."[8]

The necessity for educators to promote world government and world peace formed the basis of the World Conference of the Teaching Profession, held in Endicott, New York, August 27–30, 1946. The conference, sponsored by the National Education Association, included delegates from teachers associations in twenty-eight countries. The delegates drafted a Constitution for a World Organization of the Teaching Profession and adopted "Recommendations on the Teaching of International Understanding," which urged teachers around the world to instill in their students a sense of personal responsibility for cooperating with others for the betterment of human welfare. The recommendations covered the following eight principles:

1. People everywhere must be provided the basic needs of food, clothing, shelter, health, recreation, and security.
2. Everyone must have an equal opportunity to develop physically, intellectually, and socially.
3. The pursuit of truth and the right to express one's opinion must be unrestricted, except in cases where it interferes with others.
4. Respect for human life and the freedom of religion must be fostered.
5. No country should impose its culture upon another country.
6. The Earth's natural resources should be managed by an international organization and used for the good of all countries.
7. All people are now neighbors and mutually interdependent, and they must ensure the well-being of their neighbors.
8. The ability to ensure the security of nations, the right to self-government, and every country's opportunity for economic prosperity is contingent upon the creation of an international organization with the authority and power to promote economic cooperation and to maintain world peace.[9]

Adhering to the concepts of atomics, the conference recommended that teachers from all disciplines and grade levels work together to create a curriculum that contributed to international understanding, particularly history. But not history that emphasized wars and political struggles. Rather, the conference called for a method and curriculum for teaching history that provided students with an appreciation of how civilizations developed throughout the world. Through history, according to the conference's recommendations, "students should become familiar with the life and work of men and women of all nations who have contributed to human welfare and should learn to appreciate the spiritual and cultural heritage which is theirs." A better understanding of the past, while important, was not enough, though. The conference also emphasized the importance of teaching current events, not only to acquire immediately useful information but also to develop an intelligent interest in world affairs. The conference stressed that such information and interest in world affairs were essential for young people "if they are to fulfill their duties as citizens of their country and of the world." In addition to a new approach to history and current events, the conference outlined specific recommendations for incorporating the teaching of international understanding in modern languages, music, art, and literature, as well as revising textbooks to eliminate content "characterized by nationalistic biases and propaganda designed to promote aggressive nationalism."[10]

This sense of urgency resulted from the fact that most people believed the U.S. monopoly of the atomic bomb would end within three to five years. Thus, the alternatives were not merely peace or war. To some, like Sam Rayburn, Speaker of the House of Representatives, the alternatives were peace or the death of civilization (i.e., One World or None). The October 1945 issue of *School Life*, published by the Office of Education, reprinted Rayburn's June commencement address at the University of Maryland—a message that captured the essence of the newly unleashed atomic age. "If we do not put forth every effort to bring out an ordered world and a permanent peace," Rayburn told the graduating seniors, "we will have failed our day and generation." People must be educated in the ways of peace and concord to foster world cooperation. And to accomplish this, teachers must create civic-mindedness and develop a love of country and a commitment to democracy in their students. "We must do the job this time," he said, "in order that democracy, freedom, yea, and that civilization may survive."[11] Tracy Mygatt, representing the Campaign for World Government, took a more eloquent approach to the same message, suggesting that the choice was simple: either a world doomed to self-destruction by the "fiery breath of the split atom" or a world that recognized "man's morning vision of his destiny as an authentic potentiality."[12]

After witnessing the 1946 atomic test at Bikini Island, Lyle Ashby, assistant editor of the *Journal of the National Education Association*, expressed the alternatives this way: "Not until the military power of nations is given over to international organization can the world have or even expect to have security. The alternative is continued anarchy and finally global war of fantastic scientific ferocity—sudden death to whole populations."[13] Robert Hutchins of the University of Chicago expressed the same dire warnings, commenting in *The Educational Forum*, "The alternatives before us are no longer peace or war; they are peace or the death of civilization." Even if the world prevented war, Hutchins maintained, its economic and political systems would undergo strain due to the increased leisure time and material abundance promised in an atomic-powered era. To prepare for this possible scenario, educators must emphasize the liberal arts, "educating a man's humanity, rather than indulging his individuality." In addition, world government was mandatory, and teachers must face up to the challenge of inculcating values based on international brotherhood and not self-serving nationalism. "Civilization can be saved only by a moral, intellectual, and spiritual revolution in which we are now living," he wrote. "If American education can contribute to a moral, intellectual, and spiritual revolution, then it offers a real hope of salvation to suffering humanity everywhere. If it cannot or will not contribute to this revolution, then it is irrelevant, and its fate is immaterial."[14]

In an article published in October 1946, Harold C. Hand, one of the more influential educators in the early postwar period, quoted William Benton, chairman of the board and publisher of Encyclopedia Britannica, who said, "Education is now the only instrument of survival." Hand, education professor at the University of Illinois, concurred with Benton, arguing that not only was education critical to survival; it also was imperative that educators dedicate themselves to the concept of One World. "Plans for world government must be formulated and appraised," he wrote. "The structure of world government must first be created by and in the minds of men. No Moses is likely to come down from any mountain bearing the blueprint of an ideal world state. Instead, this design will have to be constructed through the processes of proposal and debate."[15]

Hand went on to suggest that educators must no longer teach that the United States can have everything its own way, which was a rather bold statement coming so soon after the country had defeated the Axis powers in a world war. But to Hand, the key to survival in the atomic age was embracing One World and international brotherhood. His words set the tone for other educators and social commentators in these early postwar years—and still have credence in our 21st century world today:

> It is an axiom of social psychology that people who live differently think differently, believe differently, value differently, and hence conclude differently from the same set of facts. Given the extreme wide diversities in the experiences of the peoples of the earth—which we should realistically teach in our schools—it is absurd to suppose that "they" will conclude as "we" do in any but a relatively small number of instances. It is equally absurd to assume that the peoples of the earth can live together in peace on this drastically shrunken globe on any other basis except that of compromise. We must teach accordingly, not overlooking the disciplining of our students' emotions in this regard.[16]

Before this can be achieved, though, eight propositions had to be recognized:

1. Science and technology have made everyone in the world interdependent.
2. Because of this reality, most national problems are now international problems.

3. Unfortunately, "a state of almost complete anarchy" exists in the world because of peoples' allegiance to "unlimited national sovereignty."
4. This allegiance resulted in two world wars and a global economic depression in a world dominated by old technology based on steam, internal combustion, electricity, and chemistry.
5. Now we live in a atomic age, "a completely new and vastly more potent type of energy."
6. This atomic energy has the potential of bringing great benefits to the world if we can develop a new "common sense."
7. If, however, atomic energy is not properly controlled and used, "the odds are overwhelming that man will by his own hand destroy himself and his works."
8. The new "common sense" can only be obtained through understanding, which places the burden on everyone who educates.[17]

Elaborating on these propositions, which are relevant even today, Hand stressed that everyone must realize his or her future is dependent upon people in countries around the world. Moreover, this realization must be inculcated into the minds and the emotions of children, youth, and adults. This includes a clear understanding of what he called "the international anarchy" inherited from the past. "The hurtful things which nation-states are now legally entitled to do—and *do* do—to one another with no thought except for their own short-run benefits should be detailed, documented, and appraised," he wrote. He went on to call for the study of past and present consequences of this international anarchy in terms of economic, political, and social impact on people. He then called into question the devil theory of war, which argues that political and social crises result from the deliberate actions of evil or misguided leaders. In his view, this theory merely provides the victors of war a sense of innocence—after all, they defeated "the devil," as in Adolf Hitler. Unless, and until, everyone accepts the fact that wars are a direct result of international anarchy, Hand argued, no progress can be made to prevent future wars.[18]

Nathaniel Peffer, professor of international relations at Columbia University, also examined the political implications of the atomic age, which he defined as the issue of war and peace. "Atomic energy may one day, when applied to processes of production and transportation, change the nature of civilization and society," he wrote, "but that possibility is contingent on atomic energy not having first laid civilization and society in ruins through its use in bombs." Thus, the critical question that had to be asked was this: Can we prevent another war, or at least prohibit the use of atomic weapons in war? To this end, the only alternative was between world government, either as a single entity or as a closely-bound federation, and the United Nations, an international body with limited power for the preservation of world peace. The political realities of the postwar, Peffer suggested, made the first alternative impossible. Rising nationalism and the division of the world into democratic and communist camps prevented any hope for One World. The only practical choice for Peffer was an international organization like the United Nations, although he held out the promise that the U.N. would eventually evolve into a more extensive world government.[19]

Herbert Abraham of the U.S. Department of State also heralded this view, arguing that the United Nations presented the best opportunity to establish a world order, even as the United States had taken a leadership position because of its recognition of the urgency of its role in the atomic age. "The revelation that atomic energy has been harnessed in weapons of annihilation," he wrote, "made vivid to the people of this country the weight of their responsibility. Mankind, it is felt, has been given not so much another opportunity as a last chance." Abraham concluded that the atom "is a matter of 'Death—or Life Abundant.' There is no middle ground. It's one thing, or the other. Either we control it, or we don't. If we can't control it, it is extermination." On the other hand, Abraham argued, if the world can control the atom, the future offered "fantastic abundance." But to achieve this positive outcome, schools must be decisive in their teaching of international control through a strong United Nations.[20]

As relations between the United States and the Soviet Union became more tense, educational journals intensified their efforts to convince teachers that their expanded mission was to help their students safeguard democracy. The new generation must be equipped with the requisite skills for preserving the American way of life and protecting world peace. Authoritarianism must be replaced with creativity and free-thinking. "If the child, all through school, has learned to do only what he is told, when he is told, and merely because he is told," wrote Merril Bush, education professor at Temple University in Philadelphia, "he is not likely, as an adult, to be the responsible, self-disciplined, creative, and participating citizen that a democracy demands. He is, rather, prepared to be the sheeplike follower of a dictator, and he must be kept in line by a Gestapo."[21]

Bush echoed many of his educational colleagues by focusing his comments on the ultimate choice confronting educators, government, and the world: either to create a world organization that works, or to accept the annihilation of civilization "in a gigantic holocaust which will mean the end of those ideals that typify the American way of life." And people no longer had the luxury of time to make its decision because the atomic bomb had literally blasted time away. "We stand at the dawn of a new era," Bush wrote, "which may mean a change in our ways of living as great as man's discovery of the control and use of fire. We now have within our reach the means of eliminating the scourges of famine, pestilence, and want for all men, everywhere, or, of annihilating man himself. The atomic bomb is here to stay; the question is, are we?"[22]

The American Scholar attempted to answer this question in its Spring 1946 special issue centered around the theme of "Life with the Atom." Among the contributors was Louis Ridenour, on the staff of the radiation laboratory at the Massachusetts Institute of Technology. Ridenour, who also authored the science-fiction play "Pilot Lights of the Apocalypse," took the position that the atomic bomb held some promise at encouraging nations to end all wars, but he also warned that any attempt to keep the bomb a secret was not the answer. He stressed that physicists were not astonished by news of the atomic bomb; rather, they fully understood its scientific principles and recognized that scientists in other countries would soon be able to construct similar bombs. Secrecy only served to intensify efforts by other countries to duplicate development of the atomic bomb by the U.S., thus resulting in an arms race, Ridenour warned.[23]

Erich Kahler, author of *Man the Measure, A New Approach to History*, offered a similar note of desperation by suggesting the world could only rely upon "the clemency of fate" now that the atomic bomb was a reality. According to Kahler,

> The appearance of the atomic bomb has not created a substantially new situation, but it has nevertheless completely changed the world. The state in which mankind finds itself today is no different from what it was before Hiroshima; it is one that long ago crept upon us unobserved. The atomic bomb has simply, at a single stroke, made it acute, perceptible to the senses, visible to all eyes.... "Utopia"—the world community—is today the only real thing, and all previous "practical" and "realistic" conceptions—national sovereignty and power politics and the unrestricted pursuit of economic interests—all those, from now on, are obvious delusions.[24]

The irony in Kahler's opinion was that technology had actually transcended what he called the purely technical stage. In a sense, technology had become a new moral force as the world began its search for protection from atomic warfare. Kahler argued, in fact, that this search was reminiscent of such historic morally-based endeavors as the purification and rationalization of democratic institutions, the creation of comprehensive education, and an orientation toward the human community.[25]

Adherents to this argument abounded within the educational community, including Alonzo May, economics professor at the University of Denver. May, for example, wrote that the atomic bomb threatened to rupture the country's basic social institutions unless immediate steps were taken to overhaul them. If optimists were correct in the peacetime applications of atomic energy, the resulting full employment, consumer abundance, and increasing leisure time would disrupt economic systems worldwide. Within the educational community, May contended, the liberal arts curriculum needed to be reevaluated, particularly the social sciences. Schools must take a more integrated and international approach to such subjects as government, sociology, and economics in order to achieve the "worldwide liberation of thought." Educators must reconstruct and revitalize the social sciences "so that they will serve as a true bridge between the natural sciences and the good life."[26]

Henry Christ, a teacher at Fort Hamilton High School in New York City, offered a view of someone on the front lines of postwar education. Writing in the *Journal of the National Education Association*, Christ agreed with other educators and administrators that teachers had to reexamine their roles in the atomic age because the race between education and catastrophe was nearing the finish line. Teachers needed to mold public opinion toward international cooperation, according to Christ, or "the tireless agents of destruction will win the sweepstakes."[27]

John Starie, social studies teacher at Belmont (Ill.) High School, contended similarly that education must begin a campaign for world peace; in addition, he called on teachers to help students understand the many changes occurring in the world order, including the implications of the atomic bomb. They should help break down old cultural fears and encourage more fearless questioning of the social order, requisites for children and adolescents in an era where adaptability meant survival. The challenge confronting education, according to Starie, was threefold: to serve as a means of understanding the new atomic age; to forearm society against the "cracks that must arise in the world's social stucture" resulting from the atomic bomb; and to help cushion the individual to the shocks surely to be felt in traditional mores and folkways. He wrote,

> The average pupil in our schools today was born in a world depression. He has spent a large part of his school years in the midst of a world war. He has seen the coming of a new scientific discovery. Schools and teachers cannot be true to him unless they can immediately make it possible for him to explore

for himself the effect of all of these phenomena upon himself and his society. That is the challenge that the bombing of Hiroshima laid in the lap of American education and one that American education cannot afford to ignore.[28]

School and Society, published by the Society for the Advancement of Education, followed *The American Scholar*'s special issue in June 1946 by presenting America's teachers, like Christ and Starie, with a six-point program outlining the best approach to teach atomic topics. Developed by the Society for the Psychological Study of Social Issues, a division of the American Psychological Association, the program called for educators to make clear the real possibility of another war, and to stress that no military defense existed that could prevent what it termed "the horrors" of the atomic bomb. In addition, the program called for teaching the merits of the international control of atomic energy, the advancements of international brotherhood, civilian control of atomic energy, a moratorium on the further manufacturing of atomic bombs, and an emphasis on the potential benefits of atomic energy.[29]

In December of the same year, the University of Chicago's Quincy Wright added a new element in the "peace or else" argument, and one that would continue throughout the postwar years: fear. (Also see Chapter 3.) Civilization would be sustained not because of fear of a foreign enemy but because of fear of the atomic bomb itself. In Wright's opinion, modern conditions prevented any hope for constructive measures to ensure world peace. Stronger nations never adhere to the wishes of smaller ones; furthermore, the power of atomic warfare made any semblance of a balance of power impossible. Adding to these obstacles to world peace were political rivalries, diverse ideologies, and a feeble world opinion. In order to ensure immediate world security, Wright argued, it was necessary to grapple with the power politics of nationalism; however, permanent security required going in the opposite direction (i.e., developing a world community based on international law). The only real hope for attaining this long-term objective of world peace rested in the hands of educators. Education's role, he said, was to present ideas, not propaganda; to reinforce the belief that institutions should benefit people rather than control or dictate to them; and to encourage the concept of world citizenship. An article titled "One World and the Teaching of History," appearing in the August 23, 1947, issue of *School Life*, took up the same theme, arguing that educators must teach their students that nations have been formed by giving up sovereignty while simultaneously expanding the opportunities for their people. This approach fostered better understanding of the proposals for an effective world government. John Perkins of Boston University said it more forcefully, suggesting that scientists had won the war but social scientists must win the peace. Educators, he wrote, "must harness the complex and powerful aspirations, actions, and attitudes of men and channel them into the ways of peace." The fate of mankind rested on their success.[30]

As Perkins and others argued, elementary and secondary school teachers had to take a leadership position in securing world peace by emphasizing the values of international brotherhood and the dangers of atomic destruction. *Progressive Education* editorialized in 1946,

> Atomic energy can contribute immeasurably to man's welfare or it can destroy civilization as we know it. Whether its powers shall be harnessed for good or for evil, the adult citizens of the United States will in large measure decide. It is the task of education to bring about a realization of the issues at stake and to develop the practices of human brotherhood that alone will enable us to achieve international cooperation and peaceful progress in the atomic age.[31]

Edmund Day, writing in *The Educational Forum*, also made it clear that the future rested in the realm of mind and spirit, not body and brute power. This meant, of course, that education must shape the young minds and spirits to meet the requirements of the atomic world. An example of this was the annual American Education Week, whose 1946 theme was titled "Education for the Atomic Age." Not only did schools around the nation sponsor numerous events for students, parents, and the community, but teachers formed committees to advertise school activities and to urge parents to attend open houses, while students presented talks to civic, religious, and business groups.[32]

John W. Studebaker, U.S. Commissioner of Education, often articulated the role of teachers and schools in preparing youth for life in the atomic age. Secondary education faced challenges in foreign languages, mathematics, natural sciences, practical and fine arts, social sciences, and health education. These were the challenges for a new world, he told teachers and students at the University of Michigan in 1946, "a world characterized by complexity and change, by technology and specialization, by mass communication and swift transportation, by the impulse to greater unity and, above all, by the sheer necessity of that unity if we are to escape the catastrophe of atomic warfare with its certainty of destruction for ourselves and for civilization."[33]

At the annual National Association of Secondary School Principals convention in 1947, Studebaker again acknowledged that unleashing of atomic energy had made the choice clearly between world peace and world suicide. No one really knew whether the new atomic age would bring mate-

TO THE PATRONS, STUDENTS, AND TEACHERS OF AMERICAN SCHOOLS

THE week beginning November 10 has been designated for the twenty-sixth observance of American Education Week. It should be the occasion for all citizens to visit their schools and to give serious thought to the theme selected for this year's observance, "Education for the Atomic Age."

Atomic energy can contribute immeasurably to man's welfare, or it can destroy civilization as we know it. Whether its powers shall be harnessed for good or for evil, the adult citizens of the United States will in large measure decide. It is the task of education to bring about a realization of the issues at stake and to develop the practices of human brotherhood that alone will enable us to achieve international cooperation and peaceful progress in the atomic age. —*Harry S. Truman*

EDUCATION FOR THE ATOMIC AGE is the general theme for American Education Week this year. This school-and-community-wide week throughout the Nation is sponsored by the National Education Association, the American Legion, the National Congress of Parents and Teachers, and the U. S. Office of Education.

The schedule of subjects for discussion follows. It is hoped by the sponsors that valuable results will come in every community from the 1946 American Education Week.

Practicing Brotherhood—Sunday, November 10.
Building World Security—Monday, November 11.
Facing New Tasks—Tuesday, November 12.
Developing Better Communities—Wednesday, November 13.
Strengthening Home Life—Thursday, November 14.
Investing in Education—Friday, November 15.
Promoting Health and Safety—Saturday, November 16.

During American Education Week in 1946, with the theme "Education for the Atomic Age," schools around the nation sponsored open houses and other events for parents and the community, while students presented talks to civic, religious, and business groups.

rial and spiritual benefits to all humankind, according to Studebaker, or whether a third world war would prove more totally destructive than anyone could now imagine. Again, teachers, if they were willing to move to the front and center of the world stage, held the key. "We are not working in an abstract mental laboratory," he said, "we are working with boys and girls who are destined to grow up in a ruthlessly realistic world. We have the duty of making certain that every single one of our home communities understands this. The needs of the times demand it." Students in Skokie, Illinois, heard the ruthless truth in an open letter from a member of the Atomic Energy Commission: "The blinding flash and the gigantic mushroom over Hiroshima announced the march of man, for good or for indescribable evil, into another 'new world.' It makes the building of international peace through international cooperation so much more urgent than even it was before that there is no way of using words to define the urgency."[34]

William Carr, associate secretary of the National Education Association, introduced a "Campaign for Peace" at a 1947 NEA meeting. The campaign, aimed at teachers, consisted of three components, all designed to ensure a safe future: Operation Classroom, Operation Civic, and Operation Teamwork.[35]

The objective of Operation Classroom was to reinforce the values of international brotherhood and world peace. Teachers were encouraged not to exhibit national or racial prejudice. Rather, they should teach about the United Nations and develop a sense of world community in their students. One way to accomplish this was to frame domestic issues within an international context. Teachers must wage the peace by explaining the horrors of war, Carr told the audience. But, above all, they must handle their students delicately:

> Hopeful action must not be smothered by frustrated anxiety. Release it by teaching that peace can be achieved, by considering the steps that can and must be taken to avert war, and by recognizing the benign possibilities which can be realized thru [sic] constructive use of atomic energy. Of course, you will not thrust the horrors of war upon your children in a manner, or at a time, which would impair mental health. The alternative, however, is not a saccharine education for peace which blinks blind in the cold harsh light of reality.[36]

Teachers' responsibilities were not to be confined to the classroom, Carr argued. Under Operation Civic, they were encouraged to become involved in the community, to educate parents and others about the values of world cooperation, and to explain that "a third world war need not happen, that it must not happen, that they must not let it happen, and that they can prevent it."[37]

Finally, Operation Teamwork linked teachers in the United States with those from other countries, such as those attending the NEA-sponsored World Conference of the Teaching Profession held the previous year. Carr encouraged teachers to join and support the World Organization of the Teaching Profession, which grew out of the conference. In addition, he said teachers needed to support the principle that no nation should impose its culture upon another, that natural resources should be used for the general welfare of all people, and that science had made everyone morally responsible for each other's well-being. Additionally, students should be told the truth about the cost of World War II and the potential destructive power of another one.

Teachers could not escape their responsibility to help their students, America's children, to dispel atomic anxiety and apathy (or what might be called atomic fatalism) by promoting the peacetime uses of the atom rather than the destructive nature of the bomb. Education rapidly became, according to Ronald Lora, "a vital arm of the welfare-warfare state" and part of "the canon of national security." Educational policymakers pressed teachers to help students adhere to democratic values, oppose communism, be informed about their world and their government, and become involved. R. Will Burnett of the University of Illinois, writing in *Education*, was more blunt. He warned teachers "not to scare the daylights out of people by regaling them with the horrors of atomic destruction." Instead, they should accentuate the positive aspects of atomic energy. The psychological effect of the atomic bomb, rather than the bomb itself, may, in fact, result in the death of democracy, Burnett wrote. Teachers, therefore, should concentrate on the facts of atomic destruction rather than create an aura of fear. At the same time, they should stimulate thinking and analysis on the issues related to atomic energy control, as well as develop "alert and informed international thinking" in students.[38]

Harry Gail of Westinghouse, writing in the February 1947 issue of *Progressive Education*, said a new and critical responsibility had been imposed on educators as a result of the military applications of atomic energy. "The responsibility for survival—survival of culture, of governments, of people—of our very civilization as we know it today," he argued, "[is] as real as the atom bomb itself." In Gail's opinion, teachers had to teach for world citizenship in order to assure human progress as well as to overcome misinformation about the bomb. This included the false notions that the United States had the secret to building the atomic bomb, that the U.S. was the only nation with the materials necessary to construct the bomb, that no other nation would dare attack the U.S., that a defense would soon be found against a possible atomic attack, and that an atomic bomb would not hurt U.S. cities. These falsities only encouraged apathy, which along with fear presented the greatest dangers to world survival. Teachers must present the facts of the bomb to their students, Gail maintained, and promote the concept of One World: "Each of us must strive for a world rule of order wherein the freedoms and satisfactions of our own way of life are made secure. Each of us must dedicate himself to the urgent task of ultimate survival."[39]

The American Education Federation (the renamed Progressive Education Association) formally endorsed the One World concept at its 1947 national conference at a time when its urgency was even more pronounced. Harold Rugg, of the Teachers College at Columbia University, told conference attendees that the dream of "one peaceful and cooperative world is shattered." The world now realized that World War III was a distinct possibility, and that teachers must move with assiduousness to inform their students about the realities of the atomic age. The AEF national board went on to formally adopt a new policy endorsing the establishment of a world order "in which national sovereignty is subordinate to world authority in crucial interests affecting peace and security; an order therefore in which all weapons of war and police forces are finally under that authority; an order in which international economic coordination of trade, resources, labor, distribution and standards is practiced parallel with the best standards of individual nations ... an order in which all nations, races and religions receive equal rights; an order in which 'world citizenship' thus assumes at least equal status with national citizenship." Educators, very simply, were instructed to teach the ideal of a democratic world order ... or else.[40]

School Life, continuing its commitment to the challenges of the atomic age, published a special supplement in March 1949 dedicated to atomic energy. Willard Goslin, president of the American Association of School Administrators, followed the same theme of countless educators by contrasting the choice between peace and war—life and death—and by loading the burden onto the backs of the educational community. The world, according to Goslin, stood at a fork in the road, "one prong of which leads only to destruction, the other prong taking us over the threshold to better living for more people everywhere." What people must realize before it is too late was that a greater force existed than the atomic bomb, "a force strong enough to mold the direction of our destiny, the force of education." Education had the power to lead people toward the right choice: the use of atomic energy for peaceful applications rather than war.

The world could not afford to overlook this power. "We must have a generation who understand atomic energy and its implications for a free people," Goslin wrote. "If we get such a generation, the schools of America will have to make a major contribution to their growth and development.... Time will tell if we have been equal to the task."[41]

Educating American youth about atomic energy was discussed as well by David Lilienthal of the Atomic Energy Commission. Only if students were informed about this science, according to the AEC chairman, would they be able to determine their own destiny; more important, if they did not understand the full implications of the atom, the future of democracy would be imperiled. Teachers simply could not afford to fail in their preparation of America's younger generation.[42]

Also writing in the *School Life* supplement, Mabel Studebaker, president of the National Education Association, made clear the role of teachers, and education itself, in preserving all of civilization: "America's teachers today face a heavy assignment. No matter what subject they teach, or what grade, they cannot escape the mighty questions raised by the dust of Hiroshima. Those questions make clear the menacing gap that exists between man's control of his physical environment and his control of his emotional reactions and the social framework in which he lives." Education, Studebaker argued, is the only way to close this gap. The country, and particularly educators, must "attack, and attack again, the vast wilderness of ignorance in the field of human relationships" before the world makes additional atomic bombs."[43]

According to R. Will Burnett, Ryland Crary, and Hubert Evans, in an article titled "The Minds of Men," the first task of teachers was to help in analyzing and evaluating the myriad issues surrounding atomic energy. This might not be easy, however. They pointed out that students might express fear, apathy, or fatalism (i.e., a belief that one's own opinions are unimportant or ineffective). "The Atomic Age can become a Frankenstein monster of public apathy," they wrote, "and ineptitude become widespread and persistent." Students, wrote the authors, needed to understand the issues related to controlling atomic energy, both domestically and internationally, and the destructive implications of the atom.[44]

In an article appearing in *Education*, an Illinois high school teacher reported success in this endeavor. After teaching a unit called "What It Means to Live in the World with Atomic Energy," he wrote that his students were, indeed, more convinced after the course that war was imminent, although they did learn about the peaceful applications of the atom. One student commented, "I ... realize what kind of an effect the A-bomb would have on people if ever used steadily in war, but on the other hand, how wonderfully it could serve man in peace-time."[45]

High school students around the country, including those at Denver's East High School, promoted the concept of brotherhood by participating in school-endorsed clubs such as International Relations Clubs, which sponsored assemblies and raised money for needy families. The emphasis on the United Nations, One World, and international brotherhood slowly dissipated, however, as the country approached a new decade—a Cold War decade where the clarion call to all Americans encompassed safeguarding democracy and preparing for an atomic attack from the Soviet Union. As early as May 1946, in a commencement address at Westminster College in Fulton, Missouri, British Prime Minister Winston Churchill had declared that the Soviet Union has drawn an "iron curtain" across Eastern Europe. From June 1948 until May 1949, the United States and its allies defied the Soviet Union's blockade of Berlin, Germany, with more than 200,000 flights dropping some 4,700 tons of food and supplies into the free section of the city. The Soviet Union continually blocked U.N. efforts to control atomic development, making Americans not only acutely aware of the effects and aftereffects of the atomic bomb, but also the potential threat of atomic attacks against the United States when, not if, the Soviet Union developed its own bomb. Then, in 1949, China fell to Mao Zedong and the Communist Party, and the Soviet Union successfully exploded its own atomic bomb. For many, the only option in a world with an uncertain future and an unsafe present was to suffer from atomic anxiety or to fall into a state of apathy. Clearly, apathy, or what historian Paul Boyer has called "dulled acquiescence," became the more acceptable alternative by the end of the 1940s.[46]

Although international cooperation remained an important issue, the realities of the atomic age made the defense of democracy and the promotion of democratic ideals even more critical if the world was going to survive. Nearly five years after Hiroshima, as the country prepared to stalemate its new atomic foe, Lyman Graybeal, education professor at New York University, signaled this transition from One World to safeguarding democracy. It was still essential, according to Graybeal, that education come to grips with the world crisis because the alternatives had not changed: world devastation or human betterment. Scientists, educators, and even the clergy, Graybeal suggested, were needed to make everyone realize how easily a world catastrophe could occur. Americans and the world must not only prepare for atomic destruction but for biological and chemical weapons, as well. The choice was clear. "In a real sense," he wrote, "the future of our civilization depends on the direction education takes, not only in the distant future, but in the months immediately ahead." That direction would move more clearly toward the issue of defending democracy against the tyranny of communism. And this meant that educators' responsibilities took on the added perspective of ensuring that America's school-age children would be well prepared in this quest.[47]

The increasing call for defending democracy (or, as teachers were told, "educating for democracy") also is exemplified by the American Education Federation's reassessment of its policy of educating for One World, which had not become impossible but increasingly more difficult with the emergence of the Soviet Union as a formidable atomic threat. Writing in the AEF's journal, *Progressive Education,* John Brooks, director of New Lincoln School in New York City, argued that the AEF policy "should be sounding an alarm throughout the school system to the urgent need for finding increasingly better ways to ensure man's individual and group security by processes which will eventuate some day in a world system." According to Brooks, it was now justified to teach about the common fear of war, to stress the mutual desire for security, and the recognition that "the safety of one lies in the security of all."[48]

War or peace. Life or death. Atomic bombs or atomic energy. These were the pervasive themes found in educational journals in the late 1940s. Teachers and school administrators were inundated with instructions for preparing the new generation for the atomic age. As part of the One World doctrine, students were to respect the individual; work together; acquire a sense of family devotion and responsibility; understand the need for interdependence; become a loyal American while accepting the role of world citizenship; adopt the tenets of international brotherhood; and, most important, understand fully the dangers of the atomic bomb.

Burnett and Harold Hand amplified on these responsibilities in their article "Educational Implications of the Atomic Age."[49] From their perspective, the primary citizenship and educational responsibilities for teachers, simply stated, encompassed providing the type of leadership that stimulated analytical thinking, promoted democratic values, increased realistic understanding of the world situation ("and the present crisis"), and contributed toward moral and intellectual commitments to democratic concepts throughout the world. First and foremost, however, teachers must submerse themselves in the reality of the atomic age. Even in 1951, with the emergence of the Soviet Union as a communist threat and atomic adversary, the authors heralded what they perceived as the only real guarantee for the survival of democracy: One World. Teachers, the authors contended, must understand that all modern industrialized nations are economically interdependent; communication and transportation advancements have virtually created One World, redefining time and distance; the United States and the remainder of the world are mobilizing for war; and, of greater significance, modern warfare has made the world one (as in One World or None). "A first fact of importance," they wrote, "is that an atomic war could result in brutalization, degradement, and loss of such great numbers of human beings and resources that the complex civilization that we in America and in western Europe are familiar with would have vanished."[50] This might not happen, but it could. This possibility led to another fact of modern life: The world is rampant with fear, suspicion, distrust, and hatred. Burnett and Hand portrayed the world as one displaying provincial attitudes and committed to national sovereignty and, in their view, international lawlessness. The concepts of balance of power, treaties, and armaments no longer ensured world security, they wrote; this denouement would result only from a united world committed to cooperation and harmony:

> It is for America to issue the call for world law and government of a limited order but sufficient to maintain the peace. Until this comes about the world will lick its wounds, nurse its hatreds, build its potential for destruction, and finally destroy. A world government cannot await a world community of likeminded men. A world community of like-minded men can hardly come about except on the basis of world government that guarantees peace. World government is not the answer to all the world's problems. It is the one answer, as we see it, that will make it possible to attack the world's problems at the fundamental level. If this be true the moral responsibility of education to engage the people of this country in serious consideration of this goal and its many perplexing issues is clear.[51]

This moral responsibility constituted three specifics: to provide unbiased information about the world in the atomic age so youth could make decisions based on reality, not myth; to stimulate the development of values based on individual dignity, responsibility, freedom, and justice; and, finally, to contribute to the formation of a consensus on what is right and wrong within the context of the current world situation. Educators had to build good citizens; they had to transmit democratic values; they had to disseminate undistorted, factual information; and they had to do it quickly. "We are convinced," Burnett and Hand wrote, "that education, even at this late hour, can do much to help mankind weather what seems to us to be the most difficult and hazard-freighted crisis in all recorded history."[52]

Teaching International Understanding

A year following the atomic bombings of Hiroshima and Nagasaki, and the end of World War II, teachers from twenty-eight nations met in Endicott, New York, to discuss educational requisites for the new atomic age. Sponsored by the National Education Association, the World Conference of the Teaching Profession concentrated on the necessity for educators to promote a better understanding of the world's diversity in culture and political tenets, as well as the dire necessity for international cooperation to prevent a future atomic war.

During the conference, held August 27–30, 1946, delegates drafted a Constitution for a World Organization of the Teaching Profession, published in the October 1946 issue of The Phi Delta Kappan. *The following "Recommendations on the Teaching of International Understanding" are reprinted with permission of Phi Delta Kappa International (www.pdkintl.org). All rights reserved.*

I.

The teaching of international understanding rests upon the cultivation of ideals and the development of a sense of personal responsibility for cooperation with others in all matters affecting human welfare. It does not involve the sacrifice of national culture or national citizenship, or the subordination of one to another. It requires thorough study of world problems, including the knowledge of economic forces and historical backgrounds.

This Conference, therefore, declares that teachers should instruct the youth of all lands to act upon the following principles and should strive alone or with others, to make these principles prevail in all areas of human relationship:

1. The fundamental needs of mankind for food, clothing, shelter, health, recreation, and security should be satisfied.
2. Every human being should enjoy, without any discrimination whatsoever, equal opportunity to develop physically, intellectually, and socially.
3. The pursuit of truth and the expression of opinion should be unrestricted, except when they interfere with the rights of others.
4. Respect for human life and for the religious convictions of other peoples should be fostered.
5. No nation should impose its culture upon any other nation, since no people possesses superiority by reason of divine gift, biological factors, or historical claims.

6. The natural resources of the earth should be developed by international planning and cooperation, and should be used for the general welfare of mankind.
7. The advances of science have now made all peoples neighbors, mutually interdependent and, therefore, morally responsible for each other's well-being.
8. The security of nations, their right to self-government, their cultural enrichment, and their economic prosperity can be realized only through international cooperation in an organization powerful enough to maintain peace and to facilitate worldwide economic cooperation.

II.

To the teachers and teachers' associations of all nations, the World Conference of the Teaching Profession makes the following recommendations:

1. On Education's Broad Responsibilities

The responsibilities of education for the promotion of international understanding among the peoples of all nations is a responsibility which must be shared by every teacher whether he be a teacher of children or of young people or of adults. Likewise, the obligation devolves upon every teacher no matter what his subject is. There is need for curriculum revision aimed at more direct instruction in international affairs; but there is also need for the development of desirable international attitudes and world understanding as a by-product of other lines of instruction which are not aimed primarily toward this objective.

This Conference recommends that teachers' associations in all countries take steps to bring together competent representatives of each of the several fields of instruction on each of the several grade levels for the purpose of defining their respective opportunities to contribute to the teaching of international understanding. In this undertaking the importance of coordination among the several subject fields and grade levels must be kept constantly in mind.

2. On the Teaching of History

History as a subject in the school curriculum provides unusually rich opportunities for acquainting students with the ways of living in lands other than their own, for helping them to understand the reasons for conditions that exist in the world today, and for helping them to recognize the interdependence of peoples. But history as too commonly taught fails to realize these opportunities by giving too much attention to wars and petty political struggles.

Therefore, this Conference recommends that the curriculum and method of teaching history should be such that it will give students a knowledge of the development of civilization throughout the world. Through history, students should become familiar with the life and work of men and women of all nations who have contributed to human welfare and should learn to appreciate the spiritual and cultural heritage which is theirs. History teaching in all countries should awaken young people to a sense of their responsibilities to all mankind.

3. On the Teaching of Current History

While the history of former times is essential to a proper understanding of the present, such study is not in itself enough to equip today's citizens with the knowledge and attitudes which they must possess in order to have a sympathetic understanding of the complex and constantly changing world the future of which they will help to share. Contemporary events must be studied directly—not only for the sake of acquiring immediately useful information, but also as a means of developing a lively and intelligent interest in world affairs. Such information and such interest in world affairs are essential for young people if they are to fulfill their duties as citizens of their country and of the world.

For these reasons, this Conference recommends a place for the study of current history in the curriculum, particularly secondary schools. The teaching of current history must be objective and conducted in a constructive spirit, but controversial matters should not be ignored.

4. On the Teaching of Modern Languages

Knowledge of a modern foreign language is more than a means of direct communication. The teaching of such a language offers opportunity to introduce the student to the life and habits of peoples other than his own, and, more important still, as language reflects thinking habits and character, the serious study of a modern language is a gateway to an understanding of the spirit of the people speaking it.

Therefore, this Conference recommends that modern languages be taught in order to promote international understanding and that they should be learned so effectively that pupils will be enabled to enter into the spirit of the peoples speaking them. In making this recommendation, this Conference recognizes that each country has its own particular problems and must take into consideration its neighbors and local conditions, and this Conference is also aware of the need to guard against overcrowding the curriculum.

5. *On the Teaching of an International Language*

International understanding could vastly be advanced if one language were understood throughout the world, i.e., if one language could be agreed upon and recognized as international. Two solutions offer themselves: an artificial language or a living language. Should a living language be chosen, it would be taught in all countries without, of course, replacing national languages.

This Conference is of the opinion that a living language is preferable as an international language. However, recognition of the need for subjecting the entire question to the competent study of an internationally constituted body of experts prompts this Conference to recommend that a commission be appointed by the World Organization of the Teaching Profession. This commission should be composed chiefly of teachers with scientific linguistic training and practical experience in teaching and it should include sociologists and psychologists. The task of the commission would be (a) to compare and evaluate the various artificial languages in existence; (b) to set forth what claims to internationality various living languages have, and to clarify such conceptions as "Basic English" without, however, making a recommendation for one specific language; and (c) to express their views as to the advisability of adopting an artificial language as a first or secondary international language.

6. *On the Teaching of Music, Arts and Literature*

Music and art are international in appeal. In large measure this is also true of literature. Children should not only become acquainted with artistic reproductions, but they should also be led to understand the cultural environments and attitudes of mind of the creative artists whose work they study.

This Conference, therefore, recommends:

- a) That art education include instruction about the creators of art as well as about works of art, such instruction to include the lives of artists and their respective peoples; and that the same principles should apply in the case of music education.

b) That art study should include not only painting, but also architecture, crafts, and other creative work characteristic of a country.
c) That folk lore and folk music from foreign lands be included in the school curriculum of each country.
d) That the teaching of literature include attention to the finest examples of creative writing from many countries—in translation when necessary—and that teachers have a voice in selecting the examples to be studied.

7. *On the Improvement of Textbooks*

The need for eliminating from textbooks content characterized by nationalistic biases and propaganda designed to promote aggressive nationalism has long been recognized, but it has been very difficult to establish practical means for the solution of the problem.

It is, therefore, to be noted with approval that the Education Section of the Preparatory Commission of UNESCO has offered—in its memorandum headed "Analysis of Textbooks"—four concrete proposals designed to deal with the matter. This Conference endorses those proposals. It urges teachers and teachers' associations in all countries to assist their respective "National Cooperating Bodies" to carry out the responsibilities that will be required of them in this connection and, when, needed, to prompt those Bodies to initiate action.

This Conference further urges teachers and teachers' associations to go beyond the negative action provided for in the UNESCO proposals by taking constructive steps to have included in the textbooks used in the schools of their respective countries content which supports positively the ideals of international cooperation and world understanding.

This Conference rejects the idea of a single textbook, internationally prepared for use in all countries, as being impracticable except perhaps for a textbook on the United Nations and related international organizations. Also, in the opinion of this Conference, it might be feasible and desirable to have some internationally prepared handbooks for teachers on topics of international scope.

8. *On an International Study of Teaching Practices*

Teachers in all countries can improve their programs of education for international understanding if they know more fully about the corresponding programs and practices which prevail in lands other than their own.

This Conference, accordingly, commends the proposal for an international survey of the teaching of international understanding set forth in memorandum entitled "Promotion of International Understanding through the Schools," prepared by the Secretary of the Preparatory Commission of UNESCO. This Conference, further, calls on teachers and teachers' associations in all countries to support and cooperate with the committee proposed in that memorandum.

9. *On the Extension of Literacy*

Campaigns to combat illiteracy and to promote mass education everywhere on the face of the globe merit the active support of teachers for many reasons—one of the more important being the contribution which the extension of literacy can make to the increase of international understanding. Democratically controlled relations among nations, like the practice of democracy within a nation, can be effective only to the extent that citizens are capable of keeping themselves constantly informed on world affairs, and of contributing their views to the formation of public opinion. For these purposes the citizen's ability to read and write is a minimum requirement.

In the light of these considerations, the efforts of national and international agencies to combat illiteracy and promote mass education are commended by this Conference, which further calls upon the World Organization of the Teaching Profession to cooperate in such efforts.

10. *On the Relations Among Students of Different Countries*

The value of direct experience as a means of learning is universally recognized by educators. Although opportunities for children and young people to have experience in international relations are, by the facts of geography, more limited than are corresponding opportunities in other areas of learning, there are, nevertheless, many ways by which the values of learning by experience can be utilized in the teaching of international understanding.

Among the means which this Conference recommends for the consideration of teachers in all countries are the following:

   a) Correspondence among students in different countries.
   b) Exchange of student-prepared exhibits, including such items as letters, cards, stamps, newspaper clippings, art exhibits, and school magazines and newspapers.
   c) Encouragement of students in one country to send gifts and provide other practical help for students in other countries.

d) Provision for schools and student groups in one country to give hospitality to students from other countries during vacation.
   e) Extension of present practices for student exchange, which should include college students, secondary-school students, and possibly elementary-school students, without neglecting adult education.
   f) Vacation trips for children and young people to foreign countries to be carried out under school auspices.
   g) Establishment of youth camps, which might eventually become international parks, along the common boundaries of neighboring countries.
   h) Facilitation of visits to youth hostels by students from foreign countries.

11. *On Teacher Travel and Exchange*

The individual teacher must be internationally informed and world-minded if he is to be an effective agent for the promotion of international understanding in the classroom. To this end, teachers, as well as students, need to have more first-hand contacts with the peoples of foreign lands.

To further the international education of teachers, this Conference recommends:

   a) That ample facilities for systematic international exchange of teachers be provided.
   b) That teachers' associations play a large part in the administration of programs of teacher exchange, that governments provide such associations with funds for helping to carry out such programs, and that a committee of the World Organization of the Teaching Profession be established to cooperate with UNESCO in this field.
   c) That travel bureaus of teachers' associations in different countries be revived and expanded.
   d) That governments be requested to improve facilities for international travel by revision of passport and visa regulations and sailing permits and by aiding in reduction of rates for educational travel.

12. *On the Educational Uses of Modern Media of Mass Communication*

The press, the radio, the cinema are instruments not only of amusement and recreation but also of education. The influence of the press is limited only by the extent of literacy; the radio leaps across national boundaries to inform and inspire all who have ears to hear; the cinema teaches its lessons, wholesome or detrimental, with a power and persuasiveness beyond those of the most skilled teachers and the most highly organized educational systems. These, and other modern media of mass communication, have in the past and may in the future work either with teachers or against them in their efforts to develop international understanding.

This Conference, therefore, recommends:

- a) That teachers and teachers' organizations endeavor to enlist the public press, the radio, and the instruments of visual education as potent allies for the attainment of their purposes.
- b) That the press, radio, film, recording, and television be used within the school to dramatize and invigorate the processes of teaching and learning about the modern world and that schools be furnished with the equipment needed for this purpose.
- c) That national and international radio and television broadcasts to schools be arranged and that such programs be used to celebrate events of international significance, to render tributes to great men and women of all nations, and to create a sense of human brotherhood.
- d) That films and recordings which are truly representative of the life and culture of the various nations be produced, and that the films and recordings, with suitable adaptation in language or otherwise, be freely exchanged among schools of all nations.

13. *On the Work of a Central Office of Documentation and Exchange of Materials*

Various agencies are now in existence for the collection and exchange of publications among nations and additional work of this type is expected to be undertaken by UNESCO. The activities of such agencies are already contributing significantly to international understanding; and, in our opinion, they could contribute even more if the proper authorities in charge of them would give consideration to the following recommendations of this Conference:

- a) A central office should be set up with liaison offices in every country to collect and distribute internationally such materials as reference books, treatises, textbooks, and other publications designed to be read by children and young people, and professional aids for teachers on curriculum problems and instructional methods.
- b) Teachers should play an important part in the selection of all kinds of material. While the main task of distribution should be

entrusted to the proposed central office and the liaison offices, teachers' organizations might also serve as media of exchange.

14. On Research, Experimentation, and Evaluation

Modern educational practice is ideally based on research, experimentation, and evaluation. The need for increasing international understanding through education requires new areas of specialization and the training of new experts, with continual scientific appraisal of prevailing practices, including evaluation of methods and materials.

This Conference, therefore, recommends:

a) That research be undertaken into the basis of international understanding and, conversely, into factors detrimental to such understanding—such study to include attention to the historical, psychological, and sociological which have contributed to the success of outstanding "world citizens" of the past and present.
b) That systematic efforts be made to evaluate the effectiveness of such techniques as international exchange of students and teachers, international correspondence, and use of mass media of communication.
c) That the influence of students' home backgrounds on their international attitudes be carefully studied.
d) That the best principles and practices for the guidance of foreign students be experimentally determined.
e) That advanced international studies be emphasized in universities with special attention to investigations into methods of teaching international understanding.
f) That studies should be made as to ways and means of strengthening the international roles of universities and as to the feasibility and desirability of establishing an international university. That summer courses and institutes be established for training teachers to teach international understanding.

III.

This Conference recommends that an international committee on the teaching of international understanding be established within the World Organization of the Teaching Profession, and that until such a committee is formally established the delegates to this Conference urge the associations which they represent to take appropriate action to facilitate the international exchange of ideas and information in respect to the teaching of international understanding.

IV.

Finally, this Conference urges that each one of its members aid in giving wide publicity to the foregoing recommendations on the teaching of international understanding. We especially urge that delegates publicize the recommendations among the teachers of their respective organizations and that they take all appropriate steps to encourage the adoption of the recommendations by the schools of their respective countries.

2

Teaching for the Atomic Age

> *Unless our schools can strengthen and extend their services to include the new areas brought about by the present world crisis, our heritage of freedom may completely disappear.*—State of New Jersey Division of Civil Defense[1]

On Tuesday, August 14, 1945, at 6:57 p.m. (Eastern War Time), radio stations nationwide broadcast the announcement Americans had long anticipated: "I have received this afternoon a message from the Japanese government of the unconditional surrender of Japan." For most Americans, the immediate reaction to President Harry S. Truman's succinct statement was clearly not concern about the opening of the atomic age with the bombings of Hiroshima and Nagasaki, where some 150,000 civilians had died instantly, but rather excitement and relief that a world war that had taken more than 400,000 American lives and in excess of 60 million lives worldwide had ended. Truman's words set off celebrations coast to coast, as Americans of all ages, exhausted by war, finally had an opportunity to cheer, give thanks, release years of anguish, and, for many, take solace that their loved ones still fighting on the distant islands of the Pacific would soon come home.

As for America's children, *Life* magazine proclaimed, "American kids, fans of Flash Gordon, reacted to the news [of the atomic bombings] with peanut-butter stares which seemed to say, 'What's all the excitement?' or 'We've had it for years.'"[2] Just a few days after V-J Day, the *New Yorker* caught an early glimpse of American kids at play in the new atomic age:

> No matter about grown-ups: the children are already at home in the atomic world. For years the playground in Washington Square has resounded to the high-strung anh-anh-anh of machine guns and the long-drawn-out whine of high-velocity shells. Last Sunday morning a great advance was made. We

watched a military man of seven or eight climb onto a seesaw, gather a number of his staff around him, and explain the changed situation. "Look," he said. "I'm an atomic bomb. I just go 'boom.' Once. Like this." He raised his arms, puffed out his cheeks, jumped down from the seesaw, and went "Boom!" Then he led his army away, leaving Manhattan in ruins behind him.[3]

The following month, elementary and secondary schools opened their doors with the nation at peace for the first time in four years. Yet times had changed forever. A few weeks before school began, *Time* magazine defined the new era as "the age of atomic force," saying the atomic bomb had presented "a brutal challenge" to save itself from destruction.[4] With the reality of the destructive power of atomic energy settling in, a high school newspaper commented in its first issue that fall: "Perhaps the newly developed atomic bomb affected the sudden conclusion of the war more than any single factor. It has also brought us the realization that we are living in the age of science. If this is to be the age of atomic power, and if civilization hopes to survive, it is imperative that an enduring world government be established. The necessity of hard study while we are in high school is undiminished. It is our most valuable contribution to peace."[5]

Aaron Goff, the Cleveland Junior High School teacher credited with coining the term "atomics," made clear in a 1947 article that teachers needed to spread a little fright so that students didn't become apathetic or fatalistic about the atomic bomb. Referring to his teaching colleagues, Goff wrote, "We must exert every effort, in every classroom, on every level—most certainly on an adult as well as the secondary level—to bring man's thinking up to date. Our world is no longer merely changing; it is as dynamic as a cannonball in motion, and education must keep pace."[6]

As a first step in this process, teachers had to dispel several misconceptions, beginning with the notion that science belonged in the laboratory. Goff urged teachers to accept the fact that the atomic bomb had catapulted science from the lab into the forefront not only of American society but of modern civilization. Teachers needed to grasp the reality that teaching was no longer an isolated, compartmentalized function, and, because of this, science teachers needed to be involved on every curriculum committee to ensure students fully grasped the impact of the atomic age. Teaching social studies without understanding science was as ineffectual as teaching it without understanding English, according to Goff. Yet science must be generalized, emphasizing its broader implications rather than its technicalities. Goff wrote,

> [T]he study of atomic energy on a high-school level should be undertaken primarily for its social implications and only incidentally for its pure science

aspects. We shall soon have to realize, all of us—politicians, statesmen, teachers, parents, and students—that a new world has overtaken us. Atomic energy may be its theme song or its swan song.[7]

Another misconception was the belief that "our school, our race, our town, our state, our system, our country, are necessarily perfect and superior." Goff pointed out that scientists from the United States, England, Germany, and other nations had contributed to the success of the Manhattan Project. Moreover, America was not above reproach in many areas, such as its mores and degree of democracy. Although everyone should be proud of the nation and its institutions, this did not mean that everyone should not strive for improvement. To accomplish this, he again suggested the use of the experimental and scientific approach to transform teachers into "dynamic forces in democracy."[8]

The year after Goff's suggestions, fifty-five secondary-school teachers gathered in Boston for a regional meeting to discuss how to incorporate atomic energy into their classes. The New England School Science Council sponsored the four-day workshop, with housing and personnel provided by The American Academy of Arts and Sciences. The Atomic Energy Commission also contributed by providing publications for distribution, helping to obtain speakers, and promoting the event through newspaper coverage. The goal was for these teachers to share their experiences with their colleagues, then work together to introduce atomic information to their students. To support this goal, a committee was appointed to help arrange a joint meeting of science, social studies, language, mathematics, and other teachers to make plans for incorporating the topic of atomic energy throughout the school curriculum.[9]

Social studies teachers at Brooklyn's Prospect Heights High School exemplified the workshop's goal by introducing a one-week unit on atomic energy, which eventually became one of six basic units in the World Backgrounds course of study as mandated by the New York State Education Department. Teachers were encouraged to introduce the atomic-energy unit in the second semester of American history and to cover the basic principles of atomic energy; the economic, political, and social implications of atomic energy; positive aspects of atomic energy; the nature and effects of the atomic bomb; proposals for the international control of atomic energy; and various materials available related to atomic energy.[10]

Writing in *High Points*, Leo Weitz stressed that one of the primary objectives for the unit was to instill in students the impact of atomic energy on the future of the world. To ensure a safe future, teachers needed to show students that "effective international control of atomic weapons is necessary for the survival of civilization."[11] He then offered five lesson plans.

The first lesson plan asked this question: How serious is the destructive power of the atomic bomb? The plan had a twofold objective: to raise student interest in the problem of controlling atomic energy and to help students understand the destructive powers of the atomic bomb. As with many teachers during the early postwar years, teachers at Prospect Heights High School incorporated a wide range of reading material, including *One World or None* and John Hersey's *Hiroshima*, as well as films, including the impactful *One World or None* that graphically demonstrated how an atomic bomb could destroy cities the size of New York, Chicago, and San Francisco.

Once students gained more knowledge about the atomic bomb, they went on to the second lesson plan, which explored whether a defense against the bomb existed, as well as possibilities of maintaining the secrets of the bomb. Teachers advocated the concept of international control by proposing that nation-state control of atomic energy was inadequate. This, in turn, led to a lesson plan with the primary objective of demonstrating that the survival of civilization depended upon effective international control, followed by a fourth lesson plan that focused on the peacetime uses of atomic energy. The final lesson plan focused back on the students and what they could do about "the atomic crisis." Among suggested activities were conducting a town-hall forum on atomic energy; planning and preparing a schoolwide exhibit on atomic energy; hosting an assembly program; planning and preparing a special issue of the school newspaper on atomic energy; and, finally, establishing an Atomic Energy Council that included students, teachers, and supervisors.

Endorsing the atomics curriculum, Weitz went on to recommend that atomic energy be taught in general science, physics, chemistry, biology, and even English, where teachers would facilitate the writing of compositions, editorials, and stories about atomic energy for the school newspaper. He then encouraged teachers to have students read about the Youth Council on the Atomic Crisis formed by students at Oak Ridge (Tennessee) High School.

Oak Ridge, a city of some 75,000 residents at the end of World War II, had been unknown to most of the country, including many people in Tennessee, until the war's end. Set on 59,000 acres in the eastern part of the state, Oak Ridge had come into existence in the fall of 1942 and spring of 1943 as part of the government's secret Manhattan Project to build an atomic bomb. As part of a nationwide program—which included Los

Alamo, New Mexico (the design and research laboratory); Hanford, Washington (the site for producing plutonium); and other lesser locations—Oak Ridge's mission was to separate U-235 from U-238, necessary to create a fission chain reaction critical to building an atomic bomb.[12]

Prompted in part by their parent's involvement in the development of the atomic bomb, students at Oak Ridge High School formed the Youth Council on the Atomic Crisis in the fall of 1945, a few months after the bombings of Hiroshima and Nagasaki. The idea for the council originated following an English class discussion of Norman Cousins' *Modern Man Is Obsolete*, published initially in the August 18, 1945, issue of the *Saturday Review of Literature* and later expanded as a book. Cousins had written that whatever relief and elation Americans had as a result of victory had been tempered by what he called "primitive fear, the fear of the unknown, the fear of forces man can neither channel nor comprehend." He went on to argue that nation-states had been rendered "vestigial obstructions in the circulatory system of the world," and the need for a world government had become clear in the atomic age.[13] Formation of the council finally took shape following a presentation by Dr. Charles Coryell, an MIT physicist, who spoke at a school assembly. In a clear and concise statement, the physicist told students, "There is no defense against the atomic bomb. If another war comes along, one out of every three persons in this auditorium will die from atomic blasts."[14]

Eleven students initially asked their English teacher, Philip Kennedy, to sponsor the council, and in a very short time, the eleven grew to two hundred members. The council became recognized nationally through a Christmas editorial reprinted in newspapers and quoted by radio commentators nationwide. The editorial read, in part, "We have never known a peaceful Christmas. While the atomic bomb threatens, we fear that there can be no peace for us nor for the world.... We are alarmed that this terrible menace has not been more generally recognized. We, the youth of America, must help the people see it, or we, with them, are lost." That spring, the council published a ten-page special issue of the school newspaper, *The Oak Leaf*, devoted to the atomic crisis and mailed it to 12,000 high schools, with another 1,000 later reprinted and mailed to high schools by request. Within Oak Ridge High School, the council also pushed for the integration of atomic information into the full curriculum of classes. Soon, atomic units were introduced in social studies, literature, speech, physics, and chemistry classes. Both the physics and chemistry classes, for example, studied isotopes, the cyclotron, cosmic rays, the uranium pile, possible developments of power from atomic fission, use of radioactive by-products in medicine, and the structure of atoms—with contributions by former scientists with the Manhattan Project. Speech teachers also were enlisted to improve the speaking skills of council members, with remarkable results. By August 1946, Youth Council members had traveled to ten states, participated in numerous radio broadcasts, and written more than 1,000 letters in support of peaceful applications of atomic energy. Joe Glasgow, council president, spoke at three schools in New York City, which soon formed their own YCAC groups with committees on technical aspects of atomic energy, national legislation, the United Nations, international friction, bomb damage, and world government. Following the lead of the Oak Ridge Youth Council on Atomic Energy, schools around the country began to integrate atomic issues into their activities. Similar Youth Councils were formed in twenty-five states and included both school and citywide councils.[15]

Teachers gave their pupils in New York City, from kindergarten through junior high school, an opportunity to explore the new world order in a unit titled "A Better World," first introduced in the city's public schools in the fall of 1946. The unit—based on the atomics curriculum—was designed to develop the concept of world organization and international cooperation by incorporating these themes in such diverse subjects as social studies, music, art, health education, English, and foreign languages. The ten-point program complemented similar objectives expressed in numerous educational articles of the time. These included developing respect for the individual; understanding the importance of cooperation and working together; acquiring a sense of family devotion and responsibility; comprehending the need for interdependence in group life, community life, and world affairs; being a loyal American while also accepting the role of world citizenship; learning that nations as well as individuals need one another; understanding the importance of brotherhood and acceptance of different races, religions, and nationalities; realizing the importance of economic and social security for all people; practicing responsible freedom; and displaying the American spirit of fairness, justice, and honoring the rights of others. Teachers emphasized the important role of children in their community and country, as well as in the world.[16]

In 1947, Harley School in Rochester, New York, dedicated a week in October to learning about "Living in the Atomic Age," with teachers using copies of "Living in the Atomic Age: A Resource Unit for Teachers in Secondary School," published in the December 1946 issue of the *University of Illinois Bulletin*.[17] Written by a committee under the direction of Harold C. Hand, University of Illinois education professor, the sixty-page unit contained an orientation for teachers, providing a historical overview of atomic energy; a list of questions students want to know about nuclear energy, the atom bomb, and related matters; an extensive section on suggested learning experiences; a selected bibliography; and a description of thirty-seven requisite qualities for citizenship in the atomic age, such as knowing that nuclear weapons dwarfed all pre-atomic explosives in its power to destroy life and property, and that other nations would be able to produce their own atomic bombs within two or three years. Other qualities—which reflected the teaching objectives embedded within the atomics curriculum already becoming part of America's elementary and secondary schools in the early postwar years—included the following:

- Command of scientific terminology used in discussions of nuclear energy, the atom bomb, and related topics.
- A clear understanding that the atomic bomb is a weapon of saturation that cannot be dealt with once exploded "because of its almost unbelievable heat, terrific blast and concussion, and deadly gamma rays."
- Acceptance of the fact that the fission process is common knowledge among the leading scientists of the world; and, moreover, the United States has no monopoly on fissionable materials or the engineering skills necessary to produce an atomic bomb.
- Acceptance of the harsh reality that there is no defense against an atomic bomb, and that in the next war "atom bombs will come, not in one's as at Hiroshima, Nagasaki, and Bikini, but in hundreds; and not in man-piloted planes but in radio-controlled planes or, more likely, in the noses of rockets that can be neither seen nor heard in flight."
- A belief that for the well-being of peoples of the world, at least a limited world government is essential to enforce "true world law."
- Knowledge about the potential positive aspects of atomic energy in power, heat, and medicine.
- Acknowledgement that from one-fourth to one-half of all metropolitan residents in the United States would probably be killed or incapacitated in an atomic war.
- Acceptance of the fact that the United States would undoubtedly be the first country struck in an atomic war, and that the nation's concentration of industry, government, communications, and transportation centers makes it very vulnerable to attack.

- Realization that racial, religious, ethnic, or any other brand of bigotry, prejudice, and discrimination are "enemies of peace and hence to mankind."
- Acceptance of the fact that the atomic age, with its massive destructive power, makes it imperative that the world not be allowed to "drift into an atomic war in which, if any power survives, some national state achieves by conquest a slave-type of world integration."

The endorsement of One World, as illustrated in several of the requisite qualities, was quite evident in the resource unit. "It appears that some form of world government is the most likely to succeed in bringing lasting peace to all nations of the earth," the editors concluded. "Not a world organization which counts its members as nations but one which deals directly with individuals.... We must realize what great contributions other nationalities and races can make to our culture and how much we would gain. We must be willing to compromise, moreover, if we are to live in peace in the small neighborhood which the world has now become."

Teachers at Harley School emphasized these qualities of citizenship, as well as the concept of One World, to their students in all grades and subject-matter areas. History students, for example, presented the various viewpoints of other nations toward the control of atomic energy, including how the rest of the world regarded the United States. English classes and the school library staff organized and cataloged materials to be used and researched resources at the University of Rochester, other libraries, and government agencies. Parents and others in the community contributed books, articles, and related materials. Speakers were invited to address the students, and school assemblies on atomic topics were held. Among the guest speakers was Dr. Joseph Platt, physicist at the University of Rochester, who defined the atomic age in terms of what it means in the way of new power and weapons, and told students that other countries were now capable of developing an atomic bomb. The school also set up a Geiger counter in the front hall with a piece of radioactive material positioned nearby to make it blink and click during the week. At the end of the week, sophomores, juniors, and seniors met to discuss how the week's activities contributed to their knowledge and opinions. The school's William E. Kane summed up the students' conclusions: "We must contribute our part to the creation of a world government for this new world in which we are living. Two ideas of government seem to be in constant conflict. One is authority imposed on the individual. The other is individual liberty. We must decide for ourselves what proportions of these two ideas will produce the greatest good for the greatest number of individuals in the world."[18]

Bloom Township High School in Chicago Heights, Illinois, chose the United Nations for its week-long event in October 1948. A faculty-student committee planned the "Know Your United Nations Week," which included a "Bundle Drive" in cooperation with the Save the Children Federation. Teaching the concept of the United Nations was not restricted to one week, however. The Spanish Club showcased the customs of people living in Latin America during special assemblies; the United Nations Club raised money to help schools in Italy; and guest speakers addressed the contribution of the United Nations to world peace.

According to Harold N. Metcalf, the school's superintendent, although social studies teachers had the greatest responsibility for incorporating the United Nations into their classroom studies, teachers in all disciplines should—and must—cover the concept of and principles behind the international body. He wrote,

> Physical science teachers cannot effectively deal with atomic energy without studying its use in peace and war and its control. The effects of atomic energy on life as studied in biology leads to the subject of control. Feeding the population of the world leads into areas in which the United Nations organizations play an important part. In the fine arts, students enjoy an international language and the appearance of beauty from many nations. In mathematics the teacher helps students know the thinking of men of many lands. Illustrations might be drawn from other fields of study in which effective teaching calls for the inclusion of material on world peace. All social studies areas at whatever level are excellent laboratories not only for the teaching of principles but also for carrying out activities related to the United Nations.[19]

Alexander Feldvebel, a current events teacher, used *Our Times* and *Every Week*, two school-oriented publications, to provide up-to-date information on developments within the United Nations. He emphasized the constantly evolving nature of the organization as being critical to operating more efficiently and responding more effectively to world situations. His teaching outline included the following:

I. Background
  A. The League of Nations
    1. Organization
    2. Strengths and weaknesses
    3. Causes of its downfall
  B. Events Leading Up to the United Nations
    1. The Atlantic Charter

    2. Suggested at Teheran
    3. First draft of U.N. charter at Dumbarton Oaks
    4. San Francisco Conference
  C. Organization of the U.N. at London

II. Structure of the U.N.
  A. The General Assembly
  B. The Security Council
  C. The Economic and Social Council
  D. The Trusteeship Council
  E. The Secretariat
  F. The International Court of Justice

III. Organizations working with the U.N.

IV. Important Problems before the U.N.
  A. The veto question
  B. Control of atomic energy
  C. Disarmament
  D. U.N. military force
  E. Settlement of Korean problem
  F. Seating Chinese Communists[20]

William Holder, another social studies teacher at Bloom Township High School, used group projects, including having students construct a model United Nations and creating organizational charts. Grace Chamberlain, a citizenship teacher, focused primarily on the human element rather than the U.N.'s structure (i.e., language difficulties, simultaneous translations, personalities, etc.). And Dorothy Martin's economic geography class used *World Week* magazine to discuss current events in such countries as China, Yugoslavia, Germany, and the Soviet Union—all written with the United Nations in mind.[21]

Schools not only promoted brotherhood and the need for a strong United Nations and even One World; they also fully adopted the atomics curriculum to instill the knowledge deemed crucial for boys and girls to cope in—and survive—the atomic age. Clearly, school-age children were being shaped for their mission to safeguard democracy and civilization both in the classroom and outside the classroom.

Writing in *The Elementary School Journal*, Millard Harmon argued that children had to understand the basic concepts of the atom and related matters before they were twelve in order to handle more complex topics in high school. To help accomplish this, some two thousand students—fifteen hundred of whom were in elementary school—in Newton, Massachusetts, participated in demonstrations of the principle of heat conduction in a metal bar. Ninth-grade students, the first to be involved in the experiment on the conduction of heat, responded so positively that teachers decided to conduct the experiment, along with a similar one on insulation, with sixth-graders. This led eventually to inviting all elementary-school students in seven schools to see the demonstrations, including first-graders. The conclusion from this exercise, said Harmon, was a clear indication that elementary-school education should offer a more inclusive science program that included the teaching of atomic concepts.[22]

Freshman students at University High School in Springfield, Illinois, working in groups under the direction of their teacher, developed a unit in their common learnings class titled "What It Means to Live in the World with Atomic Energy." Primarily using library resources, including books (such as *Hiroshima*), government pamphlets, and magazines, students' topics included the atomic bomb, spies, interplanetary travel, and McCarthyism (which had captured the nation's attention because of the accusations by Senator Joe McCarthy of Wisconsin that communists had infiltrated government, media, the movie industry, education, and other American organizations). Each student had to buy a copy of *Atomic Energy, the Double Edged Sword of Science*, prepared by R. Will Burnett, which served as the only common text.[23]

Teacher Audrey Lindsey sought to improve her students' ability to use the library, plan their work, make presentations, and work better in groups. Students responded well, building a glossary of more than eighty atomic-related terms, compiling a list of atomic scientists with biographic sketches, and preparing a detailed list of topics they wanted answered, ranging from the size of an atom and definitions for electrons, molecules, and compounds, to the effects of radioactivity and ways to protect yourself against an atomic bomb attack. Students taking the unit also volunteered to make models of atoms, a cyclotron, an atomic pile, and a chain reaction; make diagrams and charts illustrating the structure and destructive power of an atomic bomb; and draw maps showing distances from the Soviet Union to U.S. cities.

Commenting on the unit's effect on him, Jerry Kirk reflected the feelings of other students, saying, "At the beginning of the unit, I think the rest of the class agreed with me in feeling that the scientific viewpoint was the only phase worth studying. As time wore on, and we actually studied the social, military, moral, and other aspects, we very definitely began to appreciate these elements, too." Nancy Working, another student, added,

"Now that I look back on the unit, I think that it helped me to realize the possibilities of the atom in war, and awakened me to the limitless possibilities of the atom in helping man to increased health, safety, and prosperity. It made me aware of our responsibility to the people of the future in the ways we use this atomic energy."[24]

The Los Angeles City School Division distributed a fifty-two-page booklet titled *Atomic Energy and You* to all the city's schools in 1953. Prepared by a committee of high school science teachers, the booklet covered atomic bomb explosions, the effects of the blast, how to measure radiation, civil defense, and personal protection. At Campus Elementary School in Ames, Iowa, students took a special unit on atomic energy in 1952 that taught such concepts as the positive nature of atomic energy, the makeup of atoms and nuclear fission, and the workings of the United Nations. In Tulare, California, Kansas City, Missouri, and hundreds of other cities across the United States, high school science students built working models of atomic reactors, cyclotrons, scintillation counters, and Van de Graaf generators. Science students at Mount Baker High School in Deming, Washington, built a model plutonium production plant, while students at Anacostia High School in Washington, D.C., exploded miniature A-bombs. In 1954, teachers at 26 high schools in the State of Washington had students check radioactivity, make radioautographs, observe the course of chemicals in plans and animals, and decontaminate surfaces affected by radiation.[25]

Suffern (New York) High School students studied atomic energy in senior science class, a required course that resulted from the success of the school's Atomic Energy Club, which had formed in the fall of 1945. The club welcomed all students to join, whether they wanted to make model cyclotrons and atom smashers, design charts and posters, write articles for the school newspaper, or serve as technicians and stage hands to help with assembly programs or the club's "Atomic Medicine Show." In 1949, students built a six-foot working model of the Van de Graaf atom smasher, which gained national attention and resulted in requests from other schools for instructions for its construction.[26]

Atomic Energy Club members broadcast a one-hour radio play, "The Peaceful Atom," over station WLNA in Peekskill, New York, in 1952. Club members also led discussion groups on the desirability or futility of using atomic weapons for strategic and tactical purposes, and hosted an annual "Cavalcade Night," an open house for parents and friends for the purpose of presenting "the atomic story" to the community. During one open house, students conducted tours through the science department, which had been

The Los Angeles City School Division published and distributed *Atomic Energy and You* to schools in 1953. The 52-page booklet, prepared by a committee of high school science teachers, was designed to help study groups in schools and in area communities "become alerted to their responsibilities in civil defense preparedness."

58   Atomics in the Classroom

Science students at Mount Baker High School in Deming, Washington, built a model showing the idea of a plutonium production plant (National Education Association).

transformed into a miniature Brookhaven National Laboratory in Upton, New York, a research site dedicated to peaceful applications of atomic energy. Tables had been set up with exhibits and speakers to explain the historical background of atomic energy, the elements of an atomic bomb, fission and fusion, and the use of isotopes in medicine, agriculture, and industry. Also on display were electroscopes, a spinthariscope, Geiger counter, supersniffer, dosimeter, and a model cloud chamber—all tools of an atomic scientist.

The teaching outline for the study of the atom at Keene (N.H.) High School included the history of atomic research; the structure of the atom; natural radioactivity; nuclear fission; high-energy imparting devices, such as the Van de Graaf generator and cyclotron; the atomic pile; isotopes; the story of the atomic bomb; artificial transmutation; radioactivity detection;

applications of atomic energy in war, medicine, power, heating, and agriculture; and the necessity for universal understanding of atomic energy. Physics classes made models and mockups of an atomic power plant, a nuclear fission cabinet and a radioactivity detector, and a chain reaction was demonstrated by using two dozen mouse traps. The art department provided a model of a hydrogen atom and furnished a chart showing nuclear fission. Students also visited the Brookhaven National Laboratory exhibit in Boston and a lecture at the Massachusetts Institute of Technology. The science fair, with numerous atomic displays, was opened to the public, and the Public Service Co. in Keene provided a display window to showcase the students' work.[27]

Ninth-grade science students at the Laboratory School of the University of Chicago began studying atomic energy in the fall of 1945, with the first unit involving the reading of newspaper and magazine articles combined with demonstrations and discussion of atomic science. In 1953, the school expanded the unit during its summer session to include students who had not yet taken ninth-grade science. Eleven boys and nine girls from fourteen different elementary schools enrolled in the course, although their formal science instruction varied from none to four years. Students investigated six general problems:

1. What an atomic explosion looks like.
2. The damage from an atomic explosion.
3. How to protect oneself from the blast, heat, and radioactivity from an atomic explosion.
4. Constructive uses of atomic energy.
5. How to use scientific principles to understand and to live with atomic energy.
6. A review of scientific discoveries contributing to the development of atomic energy.

Students viewed films of atomic tests on Bikini, on Eniwetok, and in Nevada, including *Operation Crossroads*, *Operation Greenhouse*, and *Target Nevada*. They also watched *Making Atomic Energy a Blessing*, which introduced the constructive applications of atomic energy, and *Unlocking the Atom*, which tracked the scientific discoveries leading to a chain reaction. In addition, students had a variety of reading assignments, including the booklet *ABC's of Radiation* and a paper titled "Energy from Splitting Atoms," which was followed by objective test exercises. As a means to demonstrate their progress, each student prepared an exhibit based on atomic science, from showing how actual devices worked to making posters with detailed atomic principles. The highlight for students was a trip to the University of Chicago's betatron and synchrocyclotron.[28]

As the Laboratory School illustrates, atomics had, indeed, become the norm by the 1950s. Yet students also needed to learn about the atom and civil defense outside the classroom. William Reaves of the University of Chicago, argued, for example, that education should no longer be restricted to classroom activities; rather, schools must use all of their resources—including extracurricular activities and assembly programs—to prepare students for the future. "It is generally recognized that students learn from one another in informal associations on the school ground, in the school corridors, and on their way to and from school," Reaves wrote. "The almost innumerable activities in which students engage outside the high-school classrooms provide further opportunities for education through participation in the pursuit of common interests and purposes." According to Reaves, even students who did not participate in school activities benefited indirectly because well-organized extracurricular activities contributed to a school environment "more conducive to the natural and normal development of youth."[29]

One of the most successful extracurricular activities involved The Freedom Train, planned and directed by the American Heritage Foundation, which traveled to more than 300 cities between September 1947 and January 1949 showcasing 133 historical documents, including the Bill of Rights, the Constitution, and Lincoln's Gettysburg Address. Some 50 million people—or one in every three Americans—reportedly participated in Freedom Train program activities, including Youth Days organized by cities and towns hosting the exhibit, as well as fund-raising drives to buy grain for war-torn Europe. Students visiting the Freedom Train also signed Freedom Pledges, going on record in support of democracy and in opposition to communism.[30]

The month-long "Man and the Atom" and Atomic Energy Book Exhibit, part of New York City's Golden Jubilee Exposition held during the summer of 1948, drew thousands of people, including 30,000 children. Sponsored by the Atomic Energy Commission, Westinghouse Corp., General Electric, and the New York Committee on Atomic Information, the exhibit featured presentations on both the peacetime applications of the atom and its more destructive uses, as well as displays related to atomic energy, such as a radiation detector, a giant model of an atomic nucleus, and a chain reaction based on mousetraps. Attendees also received copies of *Dagwood Splits the Atom*, a comic book that explained the principles of atomic energy, courtesy of General Electric. The book exhibit, spon-

Students at Athens (West Virginia) Junior High School and their parents attend a Ground Observer Corps awards banquet in the early 1950s, during which a U.S. Air Force officer presented "wings" to the junior observer volunteers (courtesy of Joe Friedl).

sored by the American Book Publishers Council, featured titles for Americans of all ages, including many already familiar to school-age children: Norman Cousins' *Modern Man Is Obsolete*, John Hersey's *Hiroshima*, Bernard Brodie's *The Absolute Weapon*, and Daniel Lang's *Early Tales of the Atomic Age*.[31]

The reformed Ground Observer Corps' launch of Operation Skywatch in July 1952 offered another opportunity for high school students to become actively involved in protecting their country and serving their community. With radar still in its infancy, and concerns that it would not detect low-flying enemy bombers, the GOC, which had its beginnings in World War II, created Operation Skywatch initially to attract volunteers in coastal and border states to "watch the skies" for enemy aircraft, but the program expanded quickly to include all forty-eight states. In his official statement announcing Operation Skywatch, President Truman told the nation, "[I]n this new age in which hostile forces are known to possess long-range bombers and atomic weapons, we cannot risk being caught unprepared to defend ourselves.... If an enemy should try to attack us, we will need every minute and every second of warning that our skywatchers can give us." The program's objective was for a volunteer skywatcher, if he or she detected an enemy plane approaching, to pass along the information to a control, or filter, center responsible for alerting fighter interceptors and antiaircraft crews. Although the government's ambitious plans for 14,000 observation posts and one and a half million volunteers providing round-the-clock watches never materialized, more than 750,000 volunteers across the nation did participate during the 1950s, including many school-age children.[32]

In 1952, teachers responded to the new emphasis on civil-defense education by encouraging their elementary and secondary students to attend—either through field trips or with their parents—"The Show That Could Save Your Life," as it was billed: Alert America. Sponsored by the Valley Forge Foundation in cooperation with the

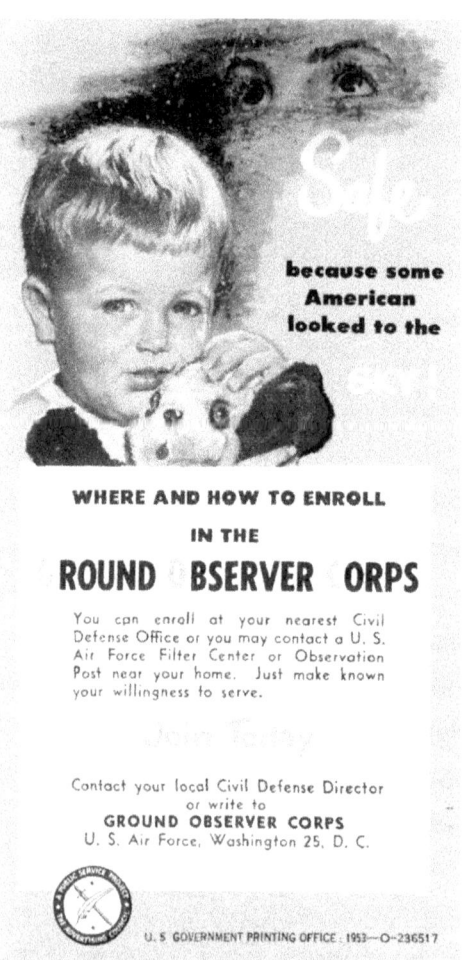

The Ground Observer Corps, reformed in the 1950s to "watch the skies" for enemy bombers, used images of children to encourage parents to volunteer. Some 750,000, mostly women, did, in fact, volunteer to do their part in safeguarding the nation.

Alert America, consisting of ten 32-foot trailers filled with portable exhibits about civil defense and the atomic bomb, traveled to more than fifty cities beginning in December 1951. The Federal Civil Defense Administration, sponsor of the exhibit, actively encouraged parents, teachers, and children to attend "The Show That Could Save Your Life." The Valley Forge Foundation published the official brochure on behalf of the Federal Civil Defense Administration.

Federal Civil Defense Administration, Alert America consisted of ten, thirty-two-foot trailers filled with portable exhibits designed to explain everything you needed to know about the atomic bomb. Over a two-year period, more than 600,000 people in fifty-six cities attended the exhibit, which included demonstrations on shelter preparation, radiation detection, and rescue procedures, and distributed "dog tags" to attendees with the person's name, address, blood type, and religion for identification in case of atomic attack. The highlight, though, was a model of a city destroyed by an atomic bomb with a spokesperson announcing that "this could be your city." According to Millard Caldwell, FCDA administrator, writing in the exhibit's official booklet, "In another war, the enemy would not try simply to kill our men in uniform, or destroy supplies and communications. His first target would be the heart of America. He would strike first at our people and our homes.... What we face is nothing less than the job of revolutionizing the thinking of the American people on national defense."[33]

In conjunction with the Alert America exhibit, schools sponsored field trips as an extracurricular activity for the purpose of preparing children for the potential dangers of the atomic age (National Archives).

A significant aspect of this job involved marketing the Alert America exhibit to America's younger generation. Part of the week-long exhibit, in fact, featured a Young Americans Day dedicated to "alerting the community to the fact that young people are ready to accept their Civil Defense responsibilities." Organizers also encouraged cities hosting the exhibit to form Young Americans committees to sponsor youth parades, rallies, quizzes, essay and poster contests, and other relevant activities. Most important, events should promote the Alert America Rallying Cry—"Alert America!"—and the Alert America Pledge:

> You can count on me to be an alert American because I want to help protect our freedoms and construct enduring peace.
>
> I WILL ... volunteer for one of the local Civil Defense services
> I WILL ... train myself and my family now in Civil Defense self-protection
> I WILL ... prepare a family shelter area and equip it with first-aid supplies
> I WILL ... prepare my home against fire and atomic attack
> I WILL ... take an active part in Civil Defense where I work
> I WILL ... take First Aid training
> I WILL ... donate my blood for our war wounded
> I WILL ... pass along the Civil Defense lessons I have learned to my friends and neighbors

The government, through such programs as the Freedom Train and Alert America exhibit, dedicated a great deal of resources—and money—to ensure that America's school-age children understood the repercussions of the atomic age and, more important, became proactive participants in it. In March 1951, as yet another example, the Federal Communications Commission reserved 209 radio channels for educational broadcasting, which allowed a unique opportunity for students to express their views on a wide range of topics. Students, in turn, responded enthusiastically across the nation, broadcasting programs over-the-air and on school networks. The Radio Guild of Proctor High School in Utica, New York, broadcast programs to elementary classes; high school students in Stamford, Connecticut, took part in weekly fifteen-minute broadcasts; students in Binghamton, New York, promoted world cooperation in radio broadcasts; and Philadelphia students presented a weekly program called *Self Help in Civil Defense*, which provided practical knowledge and skills in first aid, home nursing, and safety measures in case of an atomic attack. Beginning in 1953, with the introduction of CONELRAD, the government's emergency broadcast network, students and their parents received routine warnings about the importance of being ready for atomic attack. The public was advised to turn their AM radios to 640 or 1240 for official information.[34]

The same year, the New York State Civil Defense Commission published recommendations for teaching civil defense topics in elementary and secondary schools. "If our children and young people are to be ready to cope with national and local disasters," wrote the commission, "they must be given careful instruction in the steps to be taken for emergency protection, and careful guidance in the building of those attitudes, ideas, and understandings which will help them to become physically, mentally, and emotionally strong in the face of crises."[35] Among the recommendations were having high school students study potential problems that may occur during evacuation, participate in roundtable discussions on such topics as international cooperation and whether the atomic bomb was an overrated weapon, and learn activities designed to prevent or relieve tensions during periods of shelter confinement.[36]

Earl Peckham, director of the FCDA's Training Materials Division, writing in *School and Society*, acknowledged those teachers who had become more knowledgeable about civil defense and used this knowledge to teach the importance of civil defense to their students. He went on to compliment school districts that had integrated civil defense education into social studies, science, home economics, and other subjects. But educators could do more. "Recognizing again the excellent educational programs of communities and states promoted in certain areas of America, especially in target areas," Peckham wrote, "it is urged that there be more widespread acceptance of responsibility by educators for civil-defense education and that a more coordinated program be promoted locally, by states and through national associations."[37] He then recommended the following activities:

1. Incorporating the role of civil defense as a community program and as a student activity into social-studies classes with units on community organization, safety, and school drills.
2. Covering topics such as atomic, biological, and chemical attack in science classes.
3. Having the school-activity program reach out to provide help to the local community in preparing for an atomic attack.
4. Inviting civil defense authorities to speak with curriculum committees so teachers have more current and comprehensive information for their students.
5. Having state civil-defense agencies work more closely with state education leaders to provide educators the latest developments in civil-defense programs.
6. Ensuring that national education leaders stay abreast of civil-defense

developments and to offer their various agencies and associations in promoting civil-defense education.[38]

Unfortunately, teachers and administrators did little to allay students' anxieties about the atomic bomb by practicing "duck and cover" drills on a regular basis and incorporating atomics into the school curriculum. Whether participating in youth forums or creating posters on atomic issues, discussing the effects of an atomic blast or studying the merits of good citizenship, the unifying theme was the atomic age—an age in which two formidable adversaries, the United States and Soviet Union, maintained a precarious nuclear standoff. And although many youth maintained a strong faith in the future, suppressing any fears of atomic apocalypse, others vacillated between hope for the future and fear that there might not be one.

A Unit Outline for Teachers

Anthony Russo, a teacher in Providence, Rhode Island, served as a civil defense consultant to the city's Department of Curriculum Research in the early 1950s to develop the schools' civil defense program. His recommendations, published in The Clearing House, *emphasized the teaching of civil defense in all school subjects. Copyright 1953 from "A Unit Outline for Teachers: Civil Defense Instruction in Providence," by Anthony Russo. Reproduced by permission of Taylor & Francis Group, LLC (http://www/tandfonline.com).*

Educators agree that civil defense is a vital part of citizenship training. This article suggests in a general way the scope of an instructional unit in civil defense, considered in its broadest interpretation.

Such a unit can be taught independently of any subject (e.g., as a series of homeroom programs). However, civil defense must not be thought of as an additional subject in the curriculum. It is in fact only an additional concept to be included in teaching every subject: civil-defense instruction should permeate the entire curriculum rather than be appended as an extra subject—for which time is seldom found in the already crowded school day. Handled this way, civil defense receives the attention it deserves in context with the factors that make it necessary and with no danger of overemphasis, which might lead to unwarranted fear and anxiety.

Civil-defense material can easily be merged with the subject matter of almost every subject of the curriculum. A few suggestions are listed under Topic III of the outlined unit. Most teachers will be able to think of many more.

The outline, of course, has to be adapted to the grade level and maturity of the pupils. Many of the topics (e.g., "America's role in world affairs" and "Atomic energy in war and peace") can be treated only in sketchy fashion in the lower grades, but can be made the subjects of entire units in the secondary school, if desired. Some instruction in all the topics of the outline is needed at all grade levels, both as motivation for the schools' emergency self-protection program and as part of the schools' long-range citizenship training program. It was for these twin purposes that the outline was developed for the use of teachers in the Providence, R. I., Schools.

Outline of Instructional Unit in Civil Defense

I. Introduction

Americans need a better understanding and appreciation of a new "fact of life"—that the United States mainland for the first time in history can be attacked by a determined enemy. Modern technological advances in the science of warfare have ended for all time the relative security from enemy attack that this country has enjoyed for over 140 years. Distance is no longer an effective barrier to a determined enemy, and devastating attacks from within are now possible as never before because of these same new techniques.

Not only is the United States now vulnerable to enemy attack, but it is certain that it would be the prime target and objective in the next world-wide conflict. It was American industry and its huge productive capacity that determined the outcome of World Wars I and II; and it will be ability to outproduce the enemy that will undoubtedly win in World War III.

In modern war, industrial production facilities are major military targets. As the "arsenal of democracy," America will certainly have more of these targets than any other country.

It is a fact that although soldiers do the hard, dirty fighting on battlefields, modern wars are won in the factories and on the farms. In the last analysis, that means that civilians are as important as soldiers. Then, civilian defense must be as important as military defense. That is exactly the consensus of informed persons: real national security is not possible without an adequate civil defense. Henceforth civil defense must be a permanent and equal partner with the military in the national security organization.

The intelligent citizen of tomorrow must have a knowledge of the implications of modern scientific development in waging war, with especial reference to America's newly acquired vulnerability to attack, stem-

A Unit Outline for Teachers

ming both from these scientific developments and from our position of leadership in the world.

II. *What to Teach*

A. Self-protection
1. What to do if air-raid signal is heard—in school, at home, at play, at work—GO TO SHELTER AREA.
   a. What are the civil defense air-raid signals?
   b. What are the characteristics of a good shelter area? (Protection is needed against the effects of blast, heat, and radiation.)
2. What to do if attack occurs with no warning—DUCK AND COVER.
   a. Duck behind or under something.
   b. Cover exposed parts of the body with anything handy.
   c. Face away from windows and the light.

B. Purposes and goals of civil defense
(Degree of understanding will depend on grade level of pupils.)
1. What civil defense is. (Basically, civil defense is a plan prepared in advance to save lives and to minimize the effects of an enemy attack on the United States.)
2. Why civil defense is necessary.
   a. World Wars I and II proved that in modern times industrial production and civilian morale have as much to do with winning a war as an army and a navy. So it is expected that in World War III, the enemy will try to knock out our industries and to break down our morale (i.e., our will to fight) even before they try to defeat our Armed Forces.
   b. Modern weapons of war are weapons of mass destruction and are most effective against entire populations.
      (1) Atomic weapons and radiological warfare (against cities and people)
      (2) Biological and chemical warfare (against people, livestock, and crops)
      (3) Guided missiles (most secret of all the modern weapons)
      (4) Hydrogen bombs (still in experimental stage, but very definitely must be taken into account)
      (5) Increased destructiveness of conventional weapons (explosives and incendiaries)
   c. Enemy planes can reach every major city in the United States.

(The B-36, an intercontinental bomber, a counterpart of which we must assume that the enemy has, has a cruising range of 10,000 miles; many large American industrial cities are only about half that distance from key centers in Russia.)
  d. Experience has proved that no matter how strong military defenses are, enemy planes in large numbers will get through them (cf. London, Berlin, Hamburg, Tokyo in World War II).
  e. There is always the danger of attacks from within (e.g., fifth columnists and saboteurs).
  f. The answer: Military defense plus civil defense equals national security.

C. How civil defense will help.
  1. The individual, given all training possible, does what he can for himself in an emergency.
    a. Civil defense rests upon the principle of self-protection by the individual, extended to include mutual self-protection on the part of groups.
    b. The individual must understand the true dangers of the A-bomb and other weapons that the enemy might use.
    c. The individual must know what to do to protect himself against these dangers. (Preparedness will minimize the effects of an attack by any weapon.)
  2. The family unit, similarly trained, attacks its own problems while also contributing to the organized community effort.
    a. The family plans together what to do in an emergency.
    b. Members of the family study first aid.
    c. Members volunteer for civil defense duties.
    d. School is like a family.
  3. The community's civil defense organization works to meet its own crisis, receiving outside help if its facilities are inadequate, or contributing support to neighboring communities under organized state direction.
    a. The Providence Civil Defense Council, appointed by the Mayor, has a director, deputy director, assistant directors, and deputy assistant directors. (With one exception all members serve without pay.)
    b. Plans are being made by this council for the following emergency services: Air raid warning, warden, communication,

transportation, law enforcement, firefighting, health, welfare, engineering, rescue, evacuation, training, and mutual aid.
    c. Plans are rehearsed from time to time to see how they work and to see how they can be improved.
       (1) Operation Rhode Island—November 1951
       (2) Operation Seek Cover—May 1952
       (3) Operation Be Prepared—February 1953
  4. The State and Federal governments contribute assistance in organizational advice, overall planning, and resources.
    a. R. I. Council of Defense (appointed by the Governor)
    b. Federal Civil Defense Administration (FCDA established by Act of Congress, January 1951)

D. How school children can help in civil defense.
  1. Civil defense is everybody's job. (Why?)
  2. School children of all ages can help in civil defense by:
    a. Learning all they can about it because the more they know, the better they can help themselves, their families, their friends, and their neighbors.
    b. Learning all they can in all school subjects—reading, writing, arithmetic, social science, history, health, etc. The more good citizens our country has, the stronger it will be. The stronger our country is, the less likely it is to be attacked by an enemy.
    c. Taking home all they learn in school. By sharing their learning with their parents, they will encourage them to keep learning, too. They might urge them to join the PTA, to volunteer for civil defense work, to study first aid and home nursing. These activities will help make them more valuable citizens, too.

E. An understanding of some of the larger issues related to civil defense.
  (Degree of understanding will, of course, depend on grade level and maturity of the pupils.)
  1. Need of achieving and maintaining lasting peace among nations. (Is not peace the best defense against the atom bomb and all the other horrible weapons of modern warfare?)
  2. America's role in world affairs. (What is America doing to help achieve and maintain lasting peace in the world?)
  3. Atomic energy in war and peace.
    a. What is atomic energy?

b. Can the development of atomic energy be controlled? (American versus Russian proposals for international control of atomic energy.)
c. Use of atomic energy for military purposes forebodes the destruction of civilization as we know it.

F. Development of atomic energy for peaceful purposes has far-reaching implications for the betterment of the world.
  1. Socially (e.g., in its use by the medical sciences).
  2. Economically (e.g., in the development of low-cost power everywhere, with all that that would mean in raising standards of living).
  3. Politically (e.g., in its effect on national rivalries).

G. Need of adjusting to emergency conditions (e.g., the constant threat of war).
(Until world peace is assured, we must be prepared physically and psychologically for any attack on the United States. Civil defense contributes to both physical and psychological preparedness. Preparedness in itself will deter a potential enemy from attacking us, since an enemy who knows that his victim is ready for any emergency will be more likely to avoid open conflict on the battlefield.)

III. *Integration of civil defense instruction with regular subjects of the curriculum*

(A few typical examples are given here.)

A. English
  1. Different aspects of civil defense may be made subjects of oral and written compositions, research papers, debates, and group discussions.
  2. Civil defense pamphlets may be used as reading texts.

B. Social science, history, civics, and current events
  1. Since the need of civil defense stems directly from world political developments and unstable world conditions, greater emphasis might be laid on the study of international relations and America's role in world affairs, with its responsibilities and risks.
  2. More attention could very well be directed to the importance of polar geography to the world situation by the use of globes, air-age maps, and reports on the discovery and proposed use of the ice islands. (How is polar geography related to our national security?)

C. Sciences
   1. The sciences are inextricably involved in the problem of civil defense: atomic energy, biological and chemical agents which might be used in warfare, guided missiles are all results of scientific research. Correlation between science and civil defense instruction is perhaps the easiest of all to achieve. Many of these topics already occur in the various science curriculums; all that is needed is emphasis on their bearing to civil defense.
   2. It should be understood that all scientific discoveries may be used for either good or evil purposes—in themselves they are merely facts, neither good nor evil. Man must decide how these facts are used—whether for good or for evil.

D. Health instruction and physical education
   The school health curriculum contains much that may be applied to civil defense: First aid, home nursing, safety education, personal health, etc. Additional emphasis must now be given to these topics because they offer the fundamental techniques for self-protection and for minimizing the efforts of enemy attacks by any weapon.

E. Art
   Art activities can easily be related to civil defense: drawing of posters and building plans showing routes to shelter areas and the areas themselves, etc.

F. Special programs and assemblies
   The observance of many national holidays and other noteworthy events offers opportunities to stress civil defense relationships. A few examples follow:
   1. Washington's Birthday—present international relations in the light of Washington's advice to avoid "foreign entanglements" and the implications for civil defense.
   2. Independence Day—the contribution that civil defense can make to the preservation of our independence.
   3. Affairs Week—international tensions and civil defense.
   4. Nations Week—collective security and civil defense. Would not the need of civil defense diminish in proportion to the success of the United Nations in maintaining a stable peace?

3

Fear, Anxiety and Civil Defense

> *Civil defense is not a program of fear. Rather it is a means of overcoming fear through providing training and information, and through the development of attitudes which will help our children to be prepared for any eventuality they may be required to meet.*—New York State Civil Defense Commission[1]

President Franklin Roosevelt told Americans in 1933 they had nothing to fear but fear itself, words that energized the nation during the depths of the Great Depression. By the 1950s, at the height of the Cold War, his words gained a new, yet equally compelling, meaning: Americans, especially school-age children, had nothing to fear about the atomic bomb but fear itself. Yet government officials and educators alike believed fear that escalates to panic constituted the greatest danger to America's survival. One civil defense official even warned that fear could "produce a chain reaction more deeply destructive than any explosive known," citing an increase of more than 1400 percent in the media's use of the word "panic" between 1947 and 1952.[2]

"Fear, tempered by hope for a brave new atomic future, is part of our past and present cultural landscape," writes historian Allan Winkler.[3] During the early years of the atomic age, Americans of all ages were constantly torn between hope and fear, life and death—first, from the destructive power of the atomic bomb, then from its fallout. Historian Michael Yavenditti has described the mood of the broad population at the opening of the postwar era as one of conflicting feelings of awe, fear, satisfaction, hope, uncertainty, and bewilderment, despite the initial support for the decision to drop the atomic bomb.[4] George Hopkins, writing on the American conscience during World War II, has suggested that most

Americans believed Japan deserved the massive bombings launched in the spring of 1945—the fire bombings of Tokyo that killed some 100,000 men, women, and children—because of its misdeeds and because of racial bias against the Japanese. The atomic bombings, in his view, were initially viewed as extensions of these earlier fire bombings. "The typical American reaction to the war," Hopkins argues, "was to eradicate the threat which originally produced the war, to annihilate the enemy. Once war began, America tended to justify its actions in universal terms and pursue its goals with idealistic zeal. There was no limitation in the American way of fighting World War II. If America were to fight at all, it must fight all out."[5]

The justification for the use of the atomic bomb coexisted with more horrific images of the mushroom cloud, according to historian Peter Hales, who has suggested that Americans were torn between images of the atomic bomb as gothic horror and majestic beauty. In fact, Hales has suggested, two atomic cultures developed: one, a "marginalized group, a dissenting intellectuals' subculture" focused on the horrors and destructive power of the bomb; the other, the dominant culture, embracing the power of atomic energy, "leaving man (especially American man) to act out a predestined role in which no ultimate responsibility need to be taken."[6]

The anxiety emanating from these conflicting attitudes toward the bomb also has been well documented by historians Paul Boyer and William Graebner. The immediate response following Hiroshima, in Boyer's view, was one of confusion and disorientation as Americans attempted to confront their emotions and articulate their reactions to this "psychic event of almost unprecedented proportions." The first mandate for the American people during the early months of the atomic age was to get a grip on themselves, work together, and try to avoid future annihilation.[7] For Graebner, the country wavered between a culture of ambivalence and a culture of anxiety as Americans saw themselves living in an uncertain present with a portentous future. "Americans of the forties," Graebner writes, "lived their lives in a present that was every bit as uncertain as the depression-ridden past they were fleeing and the atomic-age future they feared."[8] For school-age children, this fear became pronounced in the early postwar years. A 1951 survey of 10,000 high-school students, for example, found that forty-seven percent of respondents worried about the possibility of atomic warfare, with thirty-six percent concerned particularly about the effects of radiation.[9]

According to historian Andrew Grossman, the Truman administration was concerned that this fear might result in what it termed "nuclear terror," which could ultimately result in Americans becoming apathetic, isolationist, or, even more alarming, supportive of atomic energy being placed under the control of an international organization—the essence of One World and the antithesis of Truman's foreign policy objectives.[10] To combat this possibility, the federal government adopted a domestic marketing campaign combined with a program for civil defense education that sought to transform the effects of nuclear weapons into those of conventional weapons, which would help reassure Americans that they and the nation would survive an atomic attack. Although initiated in the late 1940s, the government's efforts culminated with the establishment of the Federal Civil Defense Administration in 1951. Among the FCDA's objectives, according to Grossman, were managing fear and preventing panic, primarily through the publication and dissemination of more than 475 million pieces of literature by the end of the decade, with forty percent of this literature considered "home kits" aimed at parents and their children. Moreover, writes Grossman, "the federal government through the FCDA sought to contain parental fear of nuclear war by portraying the public school system as a center of security for their children."[11]

The school system may have been a center of security, but teachers still had to help students overcome their fears, according to Urban Fleege, director of the FCDA's Educational Institutions Division. Fleege outlined four types of civil defense services teachers could render: help students become self-reliant, emotionally stable, cooperative, mutually responsible, and helpful; cooperate with school plans for student safety; assist in teaching the community about civil defense; and volunteer services to local civil defense directors. Although students had been indoctrinated in the values of cooperation and teamwork under the guise of brotherhood and international relations, they also had to be self-reliant to survive an atomic war, according to Fleege. Students must not be afraid of the atomic bomb, which might result in panic; rather, they must be thoroughly prepared for it.[12]

"Sudden attack is a real threat, [although] a far less dangerous one if we are prepared for it," said Fleege. "Panic is one of the greatest dangers in an attack. Self-reliance and cooperation are responsibilities for everyone. Civil defense means personal survival." Teachers needed to emphasize to their students the real dangers of atomic attack and the importance of civil defense, while not frightening them. Moreover, Fleege encouraged teachers to train high school students in basic defense skills through regular classes such as industrial arts, physical education, health, and home economics (i.e., the atomics curriculum), as well as to expose them to operational techniques through volunteer fire fighting, rescue work, and communications. It was felt that the high school student needed to feel he or she was an integral part of civil and military defense. Teachers "must realize fear is our worst enemy," Fleege warned. But they could help allay these fears and worries by helping their students learn how to take care of themselves.[13]

The National Parent-Teacher Association (PTA) took this position in 1951 by urging its members to develop a "positive mental health program" to offset the growing fear and anxiety over the Soviet Union's atomic threat. Parents and teachers should downplay explosive headlines about this threat because children depended upon them to remain calm and confident. The PTA urged its members to help children "meet squarely" any possible emergency, including an atomic bomb.[14]

The dichotomous, irrevocable nature of the atom had permeated classrooms long before 1951, however; students actually began receiving heavy doses of atomics soon after the mushroom cloud made its initial ascent. Whether comprehensive atomic units lasting from one week to two months, or atomic issues integrated into the general education curriculum of such divergent subjects as English, history, geography, physical science, and art, stark thematic similarities were evident. Atomic discussions usually began with a baneful warning about the apocalyptic dangers of uncontrolled, unleashed atomic bombs, against which there was no defense and no protection from the effects of the blast, the heat, or the radioactivity. From the depths of fear and trepidation, students would then be lead methodically toward the light of atomic promise: the peacetime applications of atomic energy in medicine, industry, and agriculture. Ultimately, educators issued a call to arms, encouraging their students to accept the challenge of saving civilization from an adult society that had taken the world to the brink of oblivion. U.S. Atomic Energy Commission Chairman David Lilienthal's message that "this is surely no time to despair and lose hope" was repeated in elementary and secondary schools from New York to California.

Bonaro Overstreet, writing in the *Journal of the National Education Association*, suggested further that internalized fear and the atomic threat were analogous: each contributing to the growth of the other. "The present balance between life-affirming and life-denying forces is too nervously precarious to be maintained," he wrote. "The eventual swing of that balance may generally depend upon what we come to understand about fear and what we do as a consequence." The author of *Understanding Fear in Ourselves and Others* said legitimate fears contributed to the ability for human survival, but people should not fear the atomic bomb itself. Rather, they should fear authoritarian personalities, defeatism, the influence of so-called "hate-mongers," the growing sense of helplessness, human tendencies like obtuseness, and threats to free institutions.[15]

In 1953, the New York State Civil Defense Commission issued its recommendations to the state's schools to educate their students to understand the threats inherent in an atomic age but not to fear them. The commission detailed how atomics should be integrated in both elementary and secondary school courses, from history, citizenship, and world geography, to art, home economics, and science. "Nothing will so minimize fear and promote bravery among people," the commission wrote, "as will an understanding of the nature of a pending catastrophe and how to act intelligently in the situation should it develop. The program in our schools is designed to give our children security and release them from fear through knowledge, and to help them know what to do and how to do it in order to protect themselves and others if disaster strikes."[16]

Val Peterson, chief of the Federal Civil Defense Administration, explained in a 1953 article in *Collier's* magazine that the greatest danger after a nuclear attack would not be the effects of the bombs; it would be fear. To combat this fear, he recommended a series of what he called "panic stoppers," including practicing family civil defense drills at home, assembling first-aid kits, building personal fallout shelters, not participating in gossip, and, most important, having faith. "It may be faith in yourself, in your neighbors, in your leader, in your cause, your country or in God," he wrote. "In the best sense, real faith is all of them working together. But whether it is faith in yourself or in a purpose or power exceeding your own limitations, it is the ultimate solution to the ultimate weapon."[17]

Although fear became an obsession for Peterson and many people, anxiety loomed as a greater concern to others. Laurence Sears, professor of American political theory at Mills College, addressed this issue in *Educational Freedom in An Age of Anxiety: The 1953 Annual Yearbook of the John Dewey Society*. In defense of progressive educators, he issued a sharp rebuttal to the tactics of McCarthyism and the growing fear of communist subversion within society, and within the educational community. These tactics had fostered an atmosphere of panic, according to Sears, in which people were willing to give up personal freedoms because the danger appeared too formidable to overcome. "War might conceivably result in the destruction of America by Russia," he wrote. "[But] the threat of war may result in the destruction of American democracy by ourselves."[18] Sears's central thesis was that because fear is focused (i.e., the enemy is known, and a means of attack or escape can be definite), it can be overcome through knowledge. Anxiety, on the other hand, was more diffuse and, therefore, more difficult to counter. He described anxiety this way:

> There is a sense of bewilderment, of being lost, of having shelters destroyed, guards beaten down, of helplessness in the face of over-powering, if ill-defined, threats. It is when danger not only seems overwhelming but is unmanageable and the cause and cure are uncertain that anxiety results. There are times when the expectation of danger, instead of precipitating a rational fear, calls forth an overwhelming panic. It is this very sense of helplessness, of being unable to find any appropriate action to meet the danger that marks the transition from fear to anxiety.[19]

Other eras had surely experienced anxiety, Sears acknowledged, but none had encountered the uniqueness and severity of the current generation: the struggle of the United States, virtually alone and without strong allies, against the Soviet Union; the explosive power of the atomic bomb; and the subversive nature of communism. Thus, the atomic age contributed to a more acute sense of anxiety, particularly among those susceptible to emotional immaturity. H. Gordon Hullfish, editor of *Educational Freedom in An Age of Anxiety*, also warned that anxiety would be compounded if educators did not recognize that maturation was a slow process and that young people were undeveloped, helpless, and dependent. The Ohio State University education professor reminded teachers of the necessity of remaining positive as they worked to reduce anxiety and prepare the new generation for an uncertain future. "[T]hough we dare not rule out the possibility that our world may blow up as we work to improve it," wrote Hullfish, "we are entitled to have faith in the future."[20]

FCDA administrator Clara McMahon believed the best way to promote this faith in the future was through comprehensive civil defense education. "Once the resource materials on civil defense have been located and adapted to the various grade levels," wrote McMahon, "their incorporation into subject-matter areas so that students acquire the necessary understandings, skills, and attitudes is a fairly simple process." Through this education, students would develop the following attitudes:

- Responsibility for participating in a civil-defense activity.
- A desire to help others in time of need.
- A desire to become better informed about civil defense, atomic energy, world conditions, and similar current matters.
- Confidence that intelligent study and action can be an approach to the solution of problems created by world conditions.
- Open-mindedness toward the opinions of others.
- A feeling of human worth and of respect for the rights of others.

- Loyalty and steadfastness toward our democratic heritage.
- Optimism and faith in facing the future.

"We want our youth to be real citizens in every sense of the word," wrote McMahon, "a goal which can be better achieved when the teachers of America expand and extend their educational horizons to include civil-defense education."[21] In fact, a 1952 survey of more than 4,000 school systems, conducted by the National Education Association's Research Division, found that numerous changes had occurred in schools' curricula because of mobilization and defense efforts. These changes included an increase in the revision of school curricula; more emphasis on vocation work; a significant increase in physical education programs; more emphasis on social studies; and an expansion of training in first aid and safety, science, commercial offerings, and home economics.[22]

To assist administrators across the country in updating their curricula, *School Life*, published by the U.S. Office of Education, issued a checklist in 1949 to determine if schools were truly meeting the challenges of an intensifying Cold War. The most important consideration on the list was to ensure that local boards of education recognized the need for such programs—programs, which, in fact, were already part of many schools' curriculum. Teachers already understood the urgency of presenting atomic energy information in English, social studies, biology, art, geography, American government, history, mathematics and other classrooms, and schools had conducted assemblies and promoted atomic energy activities since the fall of 1945. Teachers also had used magazines, books, and newspapers, and showed films and filmstrips on the military and peacetime applications of atomic energy, with such titles as *One World or None*, *You Can Beat the A-Bomb*, and *Your Atomic World*. In many respects, the U.S. Office of Education's checklist represented more of a reminder than a new concept.[23]

In addition, high school teachers had increasingly relied on *Senior Scholastic* magazine as a weekly supplemental text providing information on current affairs and atomic energy issues. The magazine's circulation increased from 265,568 in 1945 to 309,187 by 1950—more than sixteen percent. Between 1950 and 1955, however, as the Soviet Union entered

Opposite: *Operation Atomic Vision (OAV)*, published in 1948 by the National Association of Secondary-School Principals in cooperation with the Atomic Energy Commission, was mailed free to every high school in the nation. The 96-page handbook promoted the premise that democratic survival depended upon "an informed, active public vision and enlightened leadership."

Operation Atomic Vision

New Teaching-Learning Unit for High-School Student Use

*****Operation Atomic Vision,** a teaching-learning unit for use of secondary-school students, is now off the press.

*****This** 96-page unit has been prepared by the National Association of Secondary-School Principals under the direction of Dr. Will French, Chairman of the National Association of Secondary-School Principals' Committee on Curriculum Planning and Development to encourage high schools throughout the nation to incorporate a unit on the peacetime use of atomic energy into the curriculum so that the youth of the country, and through them the adults, will understand the enormous **peacetime** potentialities of the split atom.

*****Operation Atomic Vision,** a project in community education on atomic energy for senior high schools, was prepared by Hubert M. Evans and Ryland W. Crary of Teachers College, Columbia University, and by C. Glen Hass of the Denver, Colorado, Public Schools.

***Every high-school principal of the nation has been sent **one complimentary copy** of Operation Atomic Vision. Additional copies may be purchased direct at 60 cents each with discounts for quantities.

National Association of Secondary-School Principals
1201 Sixteenth Street, N. W.
Washington 6, D. C.

—Examine **Operation Atomic Vision.**
—You will want to be a part of this nation-wide discussion project.
—**ORDER YOUR COPIES TODAY.**

the atomic age and civil defense became part of the American landscape, circulation increased by more than 120 percent, reaching 681,790 subscribers, far outpacing the 14.1 percent increase in the number of high school students. As students read the magazine's latest articles on the atomic bomb and atomic energy, teachers received instructions on how to utilize this information in their daily plans. Each issue came with a Teacher Edition containing suggested questions for class discussion, background information, and resource materials for further study.[24]

The magazine also published quizzes on atomic energy, and even offered an atomic crossword puzzle. The Scholastic Book Service provided nominally priced books on the atomic age to teachers and students. David Bradley's national bestseller and Book of the Month selection, *No Place to Hide*, an account of the atomic tests at the Bikini Atoll, was made available for just twenty-five cents. The Teacher Edition described the book as telling "how men dared the silent, colorless, odorless death from atomic radiation." High school teachers ordered thousands of these low-cost books on a variety of subjects, including atomic energy.[25]

Throughout the ensuing decade, *Senior Scholastic* continued to stress the necessity of controlling atomic energy through a world government or effective world organization. In 1953, teachers were told that the aim of an article on "Atoms for Peace" was "to help students understand that the alternative to peaceful development of atomic power for human betterment may be the destruction of civilization."[26] When Eisenhower announced the country's New Look defense based on massive retaliation in 1954, the magazine suggested that teachers ask the question, "What facts would you want to have clearly in mind before concluding that we have come to rely too heavily on atomic and hydrogen weapons?"[27]

Following *School Life*'s checklist, the Society for the Psychological Study of Social Issues, a division of the American Psychological Association, developed its own six-point program to promote the international control of atomic energy and the advancements of international brotherhood. Under the program's recommendations, teachers had to explain the dangers of the atomic bomb and the possibility of another war, plus the reality that no military defense exists to protect against "the horrors" of the bomb. The program also called for civilian control of atomic energy, a moratorium on the further manufacturing of atomic bombs, and an emphasis on the potential benefits of atomic energy.[28]

Armed with information and programs such as these promoted by the U.S. Office of Education and the Society for the Psychological Study of Social Issues, more teachers began applying their knowledge in the classroom with the aid of textbooks and supplemental materials. A study of forty-seven textbooks for junior and senior high schools published between 1945 and 1947 found increased mention of atomic-energy issues, with concentration on the bomb and international control.[29]

In 1948, *Operation Atomic Vision (OAV)*, a "teaching-learning unit" published by the National Association of Secondary-School Principals in cooperation with the Atomic Energy Commission, was mailed free to every high school in the country. The 96-page booklet, based on the premise that democratic survival depended upon "an informed, active public vision and enlightened leadership," presented both the positive aspects of atomic energy, such as applications in science and medicine, and the destructive power of atomic energy, as witnessed in the atomic bombings of Hiroshima and Nagasaki. According to authors Ryland Crary, Hubert Evans, and C. Glen Hass, *OAV* was designed to raise the level of understanding and concern about atomic energy; encourage students to help those in the community understand atomic energy; build confidence in the democratic process; and provide schools with more up-to-date information for their students. In addition, the hope was that the booklet would dissuade students from succumbing to internalized, irrational fear, as well as to overcome the lethargy and apathy toward atomic issues that had begun to settle in.[30]

The authors introduced *OAV* with an acknowledgment that students might dread the words "atomic energy" because of their fear of another Hiroshima—a fear that "will paralyze constructive action and drive us in desperation to choose destruction." To overcome this fear, students needed to understand the positive side of atomic energy, which "can be controlled and directed toward the building of a new world."[31]

In essence, OVA was a compilation of writings on all aspects of atomic energy. It featured excerpts from magazine and journal articles, books, speeches (e.g., the Baruch Proposals presented at the United Nations, which called for the establishment of an international agency to control the development of atomic energy); and legislation (e.g., the Atomic Energy Act of 1946). Among the writings deemed important for high school students were John Hersey's book, *Hiroshima*, which focused on survivors of the atomic bombing; an article by Robert Oppenheimer, technical director of the Manhattan Project, who wrote, "[I]t is apparent that a 'preventive war' is not a solution... . If we engaged in a 'preventive war,' we would commit suicide while we attempted murder"; and an article by Robert M. Hutchins, one of America's leading educators and chancellor of the University of Chicago, who wrote in 1947, "Any realistic appraisal of the status of atomic energy today, militarily speaking, must be predicated on two simple propositions: 1. There is no secret. 2. There is no defense."[32]

Writing in the *Journal of the National Education Association*, the authors told teachers, "Let us face realistically one absolute fact now: *There is no hope for the future in the A-bomb or anything connected with it* despite its apparent usefulness at the moment."[33] (*Original emphasis.*) Because of this, schools had to assume their share of the responsibility in meeting the challenge of protecting the democratic way of life, even life itself. If successful, youth would become more optimistic and better informed.[34]

Crary and Evans, writing later in an article titled "Atomic Education: A Continuing Challenge," published in *Teachers College Record*, even suggested that learning about atomic energy was perhaps the most important classroom endeavor. In this way, they argued, students could control their fears about impending disaster by obtaining knowledge about atomic issues. "Fear can induce irrational adjustments," they wrote. "[I]t can result in a refusal to confront facts; it can impel toward escape in reckless hedonism. But knowledge of fearful things is not necessarily compounded of unreasoned terror. There are awful facts of total war—the atomic bomb is one of them. One bomb at Hiroshima killed more than did the entire blitz over England. To be ignorant of or to forget this fact is to be ignorant of the hour of history in which you reside. There is still no defense, and atomic bombs have not diminished in potential fury since Nagasaki."[35]

Evans and Crary encouraged teachers to offer these "awful facts," but not to dwell on the bomb's horror, which would only contribute to fear. Rather, teachers should minimize emotional rhetoric and instead present a reasonable, balanced view of the atom: its promise and its possible peril. To overcome growing lethargy and apathy evident among the younger generation, *OAV* urged students to promote interest in atomic energy by forming discussion groups, writing letters to the editor, sponsoring lecture series, and forming atomic energy councils. *Operation Atomic Vision* soon became an important resource for secondary school teachers, including those at James Monroe High School in New York City.

In December 1948, the school devoted an entire issue of its biweekly *Bulletin of World Affairs*, published for the faculty, to social and scientific problems related to atomic energy, including its production and prospects for effective international control. Copies of the *Bulletin* were distributed to all homerooms, as well as to social studies, English, and science classes, and used by teachers to promote discussions about atomic energy, the atomic bomb, and the difficulties of living in the atomic age. In addition to setting up exhibits, students watched movies, including the March of Time's *Atomic Power*, and read and discussed John Hersey's *Hiroshima*, David Bradley's *No Place to Hide*, and Norman Cousins's *Modern Man Is Obsolete*, among other books.[36]

Ninth-grade students at William Cullen Bryant High School in New York City, for example, turned a two-week unit on atomic energy into two months, with the class divided into an art committee, which built a clay model of an atomic bomb (with all parts labeled) and a miniature version of the Oak Ridge, Tennessee, atomic installation; a newspaper committee, which planned and wrote articles; a play committee and story committee, which provided readings; and a scrapbook committee. In addition to their classroom teacher, teachers in other disciplines contributed to the unit. English teachers assisted in the writing of papers; mathematics teachers helped to calculate the number of electrons, protons, and neutrons in various elements; and social studies teachers offered background information on the current atomic energy situation in national and world affairs. Scientific principles also were reinforced through class participation. Students built an atomic model using multi-colored rubber balls. Students were then shown how to build complex nuclei by using plastic tubing. Perhaps more fun was dramatizing a chemical reaction by throwing the balls from one student to another, and moving pupils around the room to demonstrate the meaning of fission, and the concept of motion and space within the atom.[37]

The FCDA's Civil Defense Education Project contributed to atomics by publishing an array of civil defense skits or playlets to be performed by students, with titles such as "Let's Plan What to Do Now," "Operation Home Shelter," "Operation Family Car," and "Until the Doctor Comes." In a playlet titled "Skit for Planning a Home Shelter Area," Judy, a sixth-grade student, and her brother, Tommy, a fifth-grade student, explain to their father the drills conducted at their school:

> DADDY: A drill, what are you talking about, Judy? You mean you had another fire drill today? You had one yesterday.
> JUDY: No, Daddy, it wasn't a fire drill, but it was just as important. This was a shelter drill.
> DADDY: A shelter drill—what is that?
> JUDY: Oh, Daddy! In our town, we have two civil defense signals. One is the alert signal which is a steady blast of 3 to 5 minutes. This means that enemy planes are headed our way and that we should turn on our radio to 640 or 1240 on the AM dial to find out what to do.
> TOMMY: In some towns, people will be asked to evacuate their homes and schools and go to a safe place, but in our town we're supposed to turn on our radios

86   Atomics in the Classroom

to the CONELRAD stations for information, and when we hear the signal of short blasts lasting for 3 minutes, we're supposed to take shelter.[38]

Students performed these playlets and similar atomic-oriented presentations at both elementary and secondary schools. School systems throughout the country, in fact, working in cooperation with state departments, universities, and state and local civil defense organizations, rec-

ommended these types of presentations, as well as a variety of related programs, in an effort to help teachers minimize panic and hysteria in the event of a real nuclear crisis. Idaho school teachers, for example, participated in an intensive training course on atomic energy at the University of Idaho in the summer of 1953; secondary school teachers in Oregon attended a similar program at Oregon State College; and teachers in Ohio, South Carolina, Georgia, and other states received training on how to teach both the peaceful uses and the destructive uses of atomic energy.[39] Suggestions reflected the belief that knowledge mitigated fear. Thus, students needed to learn about the construction and operation of atomic bombs, as well as contamination, radiation, and fire prevention in science classes; learn first-aid skills in health classes; increase their self-sufficiency and independence from modern conveniences by learning about frontier living in social studies classes; and develop the skills to sew, use tools, cook, and grow food in practical art classes.[40]

Teachers should not be alone in these efforts, however, according to S. Mary Amatora, a psychology professor at St. Francis College. In an article titled "Emotional Stability of Children in the Atomic Age," she encouraged both educators and parents to assist children in making the adjustment to the pressures of coping in the atomic age. According to Amatora, educators must recognize the gravity of the problem of coming of age during these times of international tensions and atomic stalemate. "If the child can learn to be afraid, to be anxious, to worry, and to fear, then he can also learn the proper control of these emotions if the adults with whom he associates, both at home and in the classroom, possess and exhibit good emotional control."[41]

Writing in *The Journal of Education*, L. J. Mauth of Ball State Teachers College in Muncie, Indiana, stressed that panic and hysteria were undesirable forms of response under any circumstances but could result in mass catastrophe in time of a nuclear attack. "Each teacher must assume the responsibility for reducing the possibility of such a calamity by considering conscientiously and carefully what may be done to prevent it," wrote Mauth. "It is his obligation to prepare himself and them to meet the eventualities which one day may become actualities."[42] These actualities became more ominous with the Soviet Union's testing of a hydrogen bomb in 1953 and the domestic hysteria surrounding McCarthyism and the "Red Scare."

Opposite: **State civil defense agencies, such as the Georgia Civil Defense Division, worked closely with state departments of education to provide recommendations for incorporating civil defense education into the total school curriculum.**

State civil defense agencies also responded to the intensification of the Cold War by issuing their own guidelines that stressed the need for civil defense education to be a stronger element in schools' curricula, and the need for school administrators and teachers to provide opportunities for students to participate in civil defense programs in the community. The Georgia Civil Defense Division, for example, included a wide range of activities for pupils:

1. Discuss the present emergency, the effects of atomic blast, and protection from the atomic blast.
2. Prepare charts and posters showing the important things to remember in an atomic attack.
3. Plan first-aid demonstrations.
4. Take a field trip to observe community preparedness for an emergency.
5. Discuss causes of panic and how to prevent panic.
6. Conduct a poll among adults and high school students to discover their attitudes toward civil defense.
7. View films depicting the effects of atomic explosions.
8. Invite the local civil defense director to talk to students.
9. Read President Harry Truman's Proclamation on National Emergency (December 1950) and national and state laws.
10. Collect bulletins, letters, circulars, and pamphlets dealing with emergency problems, and interview persons who have seen the result of an atomic blast.[43]

The State Council of Civil Defense in Pennsylvania emphasized the need to mitigate children's fears in a 1952 study guide for teachers titled *Civil Defense in Schools*. The guide stressed that when children hear adults discuss the probabilities of an atomic war and mass destruction, read an article or book about atomic warfare, or watch atomic-oriented television programs or movies, they become afraid. "If they voice fears or ask questions," the guide read, "encourage them to talk. By doing so, you can help the children in two ways. First, merely expressing fears helps reduce 'tensions.' Second, you can give them correct information on how to protect themselves from an atomic bomb and increase their confidence in their ability to survive an attack.... In teaching the essential facts and procedures necessary for self-protection, it is important to teach the children WITHOUT FRIGHTENING THEM"[44] (emphasis in original).

The guide provided lessons for teaching atomics beginning in kindergarten and continuing through high school, with an emphasis on teachers

3. *Fear, Anxiety and Civil Defense* 89

remaining calm and assuring to allay their students' fears. From kindergarten through third grade, teachers should frame atomic bombs within the context of more conventional warfare and disasters such as fires and even accidents resulting from automobile accidents and firecrackers. To militate against fears from atomic annihilation, educators must rename civil defense drills—where children marched to a school shelter—as air raid drills, a common term from World War II. In an effort to calm their students, teachers had to explain that games would be played in the shelter, thus creating a new form of protective playground.

From grades four through seven, students became exposed to the impact of an atomic bomb: its "knockdown power," "heat flash," and "atomic rays." Yet teachers still needed to quiet students' fears by comparing the atomic bomb to more conventional and understood situations: "We need not fear [the atomic bomb] very much. We are much more likely to have an accident at home or from an automobile than to be injured by an

From experiments in science class, to decorating bulletin boards in art class, students in the 1950s confronted the atomic age in classroom after classroom, as illustrated in a booklet published by the Georgia Civil Defense Division.

atomic bomb. We do need to know, however, what the bomb does and how to take care of ourselves. Ignorance makes fear. Fear makes panic. Panic might cause more trouble than a bomb."[45]

Suggestions also included specific areas where high school students could volunteer, based primarily on contemporary gender roles. Girls, for instance, could volunteer as nurses' aides, mobile canteen workers, mass care center workers, child care leaders, and clerical workers. Boys, on the other hand, could volunteer as rescue workers, clean-up workers, vehicle loaders, and litter bearers. Both boys and girls were encouraged to volunteer as air observers, guides, messengers, and amateur radio operators.[46]

As the 1950s came to a close, the Oregon State Civil Defense Agency, in cooperation with the Oregon State Department of Public Instruction, published *Civil Defense in Oregon Schools*, a planning and instruction guide outlining a program to help this generation make the right decisions. In the guide's introduction, Oregon Governor Robert D. Holmes wrote,

> Some may argue that instituting civil defense into school planning and curriculum is only adding more work for our overburdened educators and is of little practical value. They could not be more wrong, for what could be more improvident than spending time, money, and effort to educate our youth and then lose them because they did not know the basic tenets of survival in the nuclear age? Our youth is the precious raw material with which to build the future. It is the responsibility of government and the individual citizen to protect and nurture this raw material as it grows into adult citizenship.[47]

With the proliferation of the hydrogen bomb and intercontinental ballistic missiles, schools became even more critical to teaching about the potential threat of the Soviet Union and the necessary knowledge and skills to survive an nuclear attack. The Oregon State Civil Defense Agency stated it this way: "Schools, as a vital institution, must carry their share of the state civil defense educational program. They have the facilities at hand to provide the education and training which are essential to enable us to meet the challenges of the times."[48] In order to be effective, school administrators, teachers, and school boards needed to be fully committed to civil defense education. The state's school superintendents were advised to appoint a civil defense planning committee whose mission was to make schools compatible with local civil defense policies and procedures. Each principal would, in turn, assume responsibility for developing, organizing, and operating his or her school civil defense program. As for teachers, the guide encouraged them not only to provide their students with instruction and practice in survival techniques, but also to instill confidence and morale in their students—a responsibility they had accepted since the atomic age began.

"Civil defense education should be part of the experience of every school-age person," according to the guide. "It prepares the student to survive physical disaster and enables him to protect himself and others, serve his community, and help strengthen the Nation in time of emergency."[49] To achieve this, the Oregon State Civil Defense Agency endorsed the atomics curriculum, stressing that civil defense instruction must be part of all classes and all grades. Beginning in the primary grades, teachers should introduce the ways in which home, school, and community are essential to the American way of life. As students progress through elementary school, teachers should begin discussions about civil defense, including target areas and plans for their defense; major transportation routes and civil defense communications systems; plans to feed people during emergencies; and the effects of nuclear attack. Civil defense issues should be integrated with health, science, physical education, and social studies topics. The guide emphasized that civil defense education in the elementary school curriculum prepared children for their increased responsibilities in high school. In addition to learning about civil defense in class, children would learn how to protect themselves by participating in shelter and evacuation drills.

Oregon high schools incorporated the atomics curriculum, with civil defense topics an integral part of social studies, science, mathematics, health education, and physical education. In social studies, for example, topics included opportunities and responsibilities for civil defense participation, the identification of target areas, resources for self-protection and mutual assistance, the world situation, and each person's rights and responsibilities in a democracy. Class discussions were not enough, however. The guide also recommended after-school volunteer activities, including helping and entertaining younger children, as well as assisting the handicapped; serving as civil defense messengers; performing auxiliary duties, such as fire wardens, room wardens, first-aid workers, stretcher-bearers, home nurses, and loading zone monitors; operating amateur radio or working on the school telephone switchboard; serving meals or performing clean-up duties following mass feeding operations; and being parking lot attendants at emergency welfare centers.

In order to implement these various guides, recommendations, and mandates for dispelling students' fears and anxieties, while simultaneously encouraging them to accept their role in civil defense and being prepared to survive a nuclear war, educators became, in essence, social engineers, or change-agents. In the words of the University of Michigan's David Jenkins, social engineering was "the controlled planning and arranging of a series of events or procedures to lead to some determined result." The atomic age, with its myriad trepidations and resulting fears and anxieties, demanded such action, according to many postwar educators. Kenneth Benne, education professor at the University of Illinois, even outlined five basic democratic norms for what he called "the engineering of change." First and foremost, teachers and school administrators must work together in order to achieve the desired result. The engineering of change also must be presented as educational to students, be experimental, and be task-oriented. In a reformulation of the tenets of brotherhood, Benne argued that this approach must be anti-individualistic, "yet provide for the establishment of appropriate areas of privacy and for the development of persons as creative units of influence in our society." In other words, educators, as social engineers of youth, must discourage individualism as a threat to democracy and world survival, yet promote the values of individuality, or the belief in individual worth. Ralph White of the Central Intelligence Agency, writing in *Progressive Education*, argued strongly that the ultimate meaning of democracy was respect for the individual; however, there must be a balance between concern for individual welfare and concern for group achievement. If reinterpreted within the context of the atomic age, this social engineering was designed to develop strong individuals who, in turn, would safeguard democracy and secure the safety of the group: a principle that became more pronounced as the 1950s unfolded.[50]

Civil Defense in the Classroom

By the early 1950s, with the Soviet Union's entrance into the atomic age and the escalation of the Cold War, civil defense became an integral part of the atomics curriculum. Although many of the same topics and themes continued from the early postwar years, they were now presented within the context of being prepared in case of an atomic attack, considered a very real probability by many. In 1953, The New York State Civil Defense Commission published Civil Defense and the Schools, *guidelines for elementary and secondary school teachers to integrate civil defense into all subject areas, from citizenship, American history, and geography, to art, health education, and home economics.*

Civil Defense in the Elementary School

In the kindergarten and grades one through six, the total curriculum plays a part in the development of skills, ideas and concepts essential to

CIVIL DEFENSE
and
THE SCHOOLS

NEW YORK STATE CIVIL DEFENSE COMMISSION

Thomas E. Dewey
Governor

Lawrence Wilkinson
Chairman

Clarence R. Huebner
Director

1953

civil well-being and protection. In the areas of citizenship education, health and physical education, and in science, particularly, it is possible to bring out and emphasize civil defense content. A few suggestions as to how this may be done are included in the pages that follow.

Citizenship Education

The program in citizenship education can do much to help children become mentally alert, emotionally well balanced and skilled in the abilities and understandings necessary to effective democratic living. Such a program in action presents many opportunities for older children to share in or assume responsibility for helping younger children. This bond of responsibility, which would be of great importance in an emergency, can be developed through such everyday citizenship projects and activities as milk distribution, cafeteria assistance, safety patrols, and school and grounds beautification campaigns.

School-wide Defense Committees

As a part of the citizenship education program, some schools are organizing civil defense committees in the upper grades. The membership of these committees is made up of representatives from grades four, five, and six, usually a boy and a girl from each group, who work under the careful supervision of a committee of teachers. Parent groups are often represented, and school administrators are usually invited to attend all meetings.

These civil defense committees invite local civil defense personnel to explain the civil defense program both to the children in the upper grades and to their parents. They distribute civil defense literature to older children and to parents. They help also to plan programs which provide for:

1. Instructions to the school children on the proper action to be taken when the air-raid alarm is sounded during in-school hours.
2. Instructions to the school children concerning the proper action to be taken when on buses during the sounding of the air-raid alarm.
3. Instructions to the school children on the proper action to be taken when the air-raid alarm is sounded while they are outdoors.
4. Instructions to the school children concerning what to do immediately after hearing the explosion of a bomb and seeing a flash of brilliant light.

Cooperation in such activities helps children develop satisfactory social relationships. Children learn to be worthy group members by having opportunities to work in groups. Children can be helped to develop the attitude that they work primarily for the common good through the setting up of many school-wide projects, activities and campaigns. Experiences of this nature will help them to work effectively with others under emergency conditions.

Patriotism

A very important function of the citizenship education program is to help children develop that inner strength and fire which has its roots in love of country. Throughout the elementary school years children are given many opportunities to develop an understanding of their American heritage and to participate in patriotic activities.

Beginning with their very first year in school, children celebrate our national holidays, salute the flag, and sing simple patriotic songs. They learn something of the contributions to their country made by such famous Americans as Washington and Lincoln. As they continue through the grades they begin to study the history of their community and their country. Recordings and movies are used whenever possible to make history become alive. Through reading about and dramatizing significant events and episodes from the lives of famous persons, the children build a background of information which helps them to understand and appreciate their heritage as Americans. It is during these formative years spent in the elementary school, and through such experiences and studies as these, that children come to identify themselves with their country and are willing to work and later to fight, if necessary, for what it stands for.

Neighborhood and Community Studies

In the early years of the elementary school, the content of the citizenship education program deals with the child's immediate environment—the home, school, neighborhood and community.

There are many opportunities in these years to emphasize civil defense concepts and understandings. For example:

1. In studying the manner in which the home, neighborhood and community provide for the basic needs of life (food, clothing and shelter), spend some time in finding out how these needs would be cared for in time of emergency.
2. In studying transportation and communication in the neighborhood

and community, find out how these essential needs will be cared for in time of emergency. It might be well to stress the importance of "staying quietly at home" so as not to block vital roads with civilian traffic.
3. In studying community health, ascertain what provisions have been made to care for emergency victims, and for meeting health needs in the safety zones and shelters.
4. In studying about such community helpers as policemen and firemen, look into the roles that they would play in the event of an emergency.
5. In studying the industries of the local community, discuss the extent to which they are involved in the defense effort and how they would be kept in operation in time of emergency.

Regional Studies of the United States

The fifth grade program draws most of its content from the history and the geography of the United States. The teaching of history has as its goal the building of an understanding and an appreciation of our American heritage, resulting in the development of a firm and lasting patriotism.

In teaching the geography of our country, it is possible to emphasize civil defense through such topics and activities as:

1. The major industrial areas of the United States, bringing in plans for their defense.
2. The major transportation routes and communications centers of the United States, and their importance to civil defense planning.
3. The importance of our natural resources in the defense effort.
4. The major food producing regions of the United States, and provisions made for food distribution in time of emergency.
5. In current events, keep a bulletin board or scrapbook of news items pertaining to civil defense activities.
6. Using as much information as is available, discuss what might happen in the event of enemy attack on a major metropolitan area.

Regional Studies of Other Areas

The program in grade six deals with other areas of the world. During this year, it is possible to emphasize topics and activities, pertinent to civil defense, such as:

1. Safeguarding our country through foreign aid programs and defense alliances.
2. The general nature of the people and countries that are considered to be our enemies.
3. The general nature of the people and countries that are our allies.
4. The role of the United Nations in promoting peace.
5. United States military installations overseas—our first line of defense.
6. In current events, study and discuss clippings which deal with civil defense, with our overall defense program, and with our relationships with our allies and enemies.

The foregoing is a brief statement of the civil defense emphases which can be brought into the citizenship education program. It is not intended to be a complete coverage of all the possible aspects of civil defense education that can be included. It is hoped that local school personnel will add to this list as they see opportunities to make civil defense ideas and thinking a part of the everyday living of children in ways which will not stimulate fear or build up tensions.

Health Education and Physical Education

All phases of the health and physical education programs are important to civil defense, since these programs have as their major goals the development and maintenance of sound minds and bodies. Certain areas of these programs, however, can also be given a direct civil defense emphasis, notably health protection, safety, first aid, shelter activities, and school camping.

1. Health protection: an area in which it is possible to discuss the sanitary use of emergency cooking, drinking and toilet facilities, safeguards against infections and damage to the eyes, etc., from bomb blasts, as well as many other ways of taking care of one's own and others' health under emergency conditions.
2. Safety: an area in which it is a good idea to stress safe conduct during drills and shelter activities, precautions which might be taken to minimize fires due to bomb blasts, the safety functions of local police and fire departments in time of emergency, and the whole area of safety precautions designed to minimize personal injury during bombing.
3. First aid: an area in which simple first aid procedures, in accordance with the maturity levels of the children, could be stressed, as well as

the vital importance of having first aid supplies readily available at home and in school, and the necessity of having parents and teachers trained in the use of such supplies.
4. Physical education: an area which is devoted to helping children acquire the fundamental skills and correct knowledge of body mechanics which they would need to take care of themselves and others in time of emergency.
5. School camping: an activity which affords children an opportunity to learn the skills necessary for survival under more or less primitive conditions, such as those under which they might have to live in an emergency, i.e., outdoor cooking, preservation of foods, selection of clothing to keep warm and dry, and the construction of temporary shelters.
6. Shelter activities: suggested types of activities which may be found helpful in satisfying the need for physical activity and for relieving emotional tension during periods of confinement in shelter areas are:

Mimetics or Imaginative Play

Jack in the Box (Leader opens and closes cover of box. Children do deep knee bend when closed, and stand with "squeek" when opened. Vary the speed of openings and closings)

Jack Be Nimble (Stress height)

See-Saw

Up and Down the Scale or Elevator (Start from deep knee bend. Children rise slowly as they sing up the scale and reverse going down the scale)

Singing Games

Red Bird

Did You Ever See a Lassie? (Eliminate skipping)

Looby Loo (Eliminate skipping)

Rhythm or Rhythm Plays

Where, Oh Where, Has My Little Dog Gone?

I'm Very, Very Tall

Jolly Is the Miller (Eliminate skipping)

Oh, Dear, What Can the Matter Be? (Use movements suggested by the song)

Folk Dances
    Shoemaker's Dance
    Danish Dance of Greeting (Eliminate running steps)

Self-testing Activities
    Duck Walk (Pivot in small circle)
    Turk Stand and Sit
    Jump, Slap Heels
    Upswing
    Chinese Get-up
    Half Top Spin

Body Mechanics
    Standing, sitting, lifting, pushing, pulling, etc.

Active Games and Relays
    Cap In
    Hotball
    Follow Me
    Ducks Fly
    Poor Pussy
    Lion Hunt (Not in standard texts, but well known to most physical education personnel)
    Quiet Games
    Beast, Bird or Fish
    Buzz
    Initials
    London
    Music Box
    Prince of Paris
    Scat

The Science Program

One of the primary purposes of elementary science instruction is to help children develop the ability to think objectively and scientifically. In addition to the teaching of basic science content and understandings, the science program gives children opportunities to utilize scientific methods of thinking in working out solutions to problems. It helps them to think clearly about the nature of problems, to apply known facts, to determine needed information, and to arrive at conclusions that can be supported.

This manner of thinking can be applied with equal effectiveness in all areas of living. It might well be particularly important in dealing with problems growing out of emergencies.

Certain areas of science content can be related directly to civil defense. Among them are:

1. Conservation—which is simply good sense and yet is very important to civil defense programs. Activities which stress the civil defense implications of the conservation of food, clothing, water, personal possessions and natural resources will help to develop attitudes useful to survival in time of emergency.
2. Electricity and magnetism—which give us power, heat, light, and communications. The importance of having emergency sources of power might well be stressed, particularly in those communities where no such provision has been made.
3. Chemical change—which provides an opportunity to look into the effects of radiation on water and foods. It would be well to bring out the importance of setting up reserve supplies of canned goods and bottled water, both on an individual and a community basis.

These and many other science topics and activities can be used to help children understand that we must be ready to meet an emergency which might come at any time.

Conclusion

Nothing will so minimize fear and promote bravery among people as will an understanding of the nature of a pending catastrophe and how to act intelligently in the situation should it develop. The program in our schools is designed to give our children security and release them from fear through knowledge, and to help them know what to do and how to do it in order to protect themselves and others if disaster strikes.

Civil Defense in Secondary School

Civil defense emphasis in the secondary school program is intended to continue and expand those abilities, skills and understandings, the foundations of which were laid in the elementary school. The greater maturity of junior and senior high school youth makes it possible to introduce many new civil defense emphases in additional subject areas.

It is possible in many of these areas to bring in a great deal of content relating to civil defense. A number of suggestions as to how this may be done are included in the pages that follow. This is by no means a complete list, and it is hoped that individual teachers will find additional ways to help their students acquire the information and skills necessary for effective action in times of local and national emergencies.

Citizenship Education

A major function of the citizenship education program in every school is the development of local responsibility and initiative. This is particularly important with regard to the civil defense program on the local level. Teachers of citizenship education can do much to help pupils acquire the understandings and skills necessary for them to become effective participants in the civil defense programs of their communities.

STATE AND LOCAL COMMUNITY (GRADE 7)

In the various units of the seventh grade course, particularly in Units I, V, VI and VII, teachers are afforded opportunities to relate civil defense activities to the general objectives of citizenship education. The following are possible activities and topics growing out of selected aspects of the regular program:

1. In the school orientation program, have pupils become acquainted with the members of the school staff who have assumed specific civil defense responsibilities. Pupils might interview these staff members and report to the class on their civil defense duties.
2. In teaching about the school plant, provide for a study of the plan of the school and how safety and defense rules should be observed in the school. Point out what should be done in an emergency and how this could be applied in everyday life.
3. Study the regulations governing civil defense practice drills in preparation for an enemy attack.
4. Include the implications for civil defense in teaching the geography of the local community and the state.
5. Use community surveys to familiarize pupils with the resources of the community and with the manufacturing, transportation and communication systems. Ascertain shifts in the community's population within a 24-hour day.

6. Find out from the local civil defense director whether the community has been designated as a target or support area. In teaching the economic life, resources and culture of the state give special attention to the type of area and explain why it is classed as such.
7. In the unit on state and local government, provision could be made for a study of the structure and functions of the State Civil Defense Commission and for a study of the local civil defense organization. Indicate the expenditure for civil defense as the class studies state and community budgets.
8. When community services for health and welfare are studied, include the emergency medical and welfare services of the local civil defense organization.
9. Include, among people who can be invited to the classroom to describe community services, police, fire and other civil defense officials who can explain the broad phases of civil defense protection and safety measures provided for in the community.
10. Make sure that the home phases of civil defense are included in the study of the family group. Pupils should know their own responsibilities for self-protection at home, and also may help to make their families more civil defense conscious. A check-list or questionnaire on home preparedness against fire or explosion and on the locations of nearest shelters might be developed by the class. Each family should arrange for a common meeting place in the event of a disaster which separates them.

Our American Heritage (Grade 8)

The eighth grade course, with its major aim "to develop love of country, pride in American heroes and achievements, democratic ideals and a sense of the duties and obligations as well as the privileges of citizenship," should stimulate the enthusiasm of junior high school pupils to take an active part in civil defense.

The following are possible activities and topics related to the units mentioned:

1. Have pupils make illustrated maps of the United States showing the general locations of atomic energy installations and of areas in which important defense materials are produced. Show pictures, slides or filmstrips about American industrial production for defense purposes.
2. Stress the importance of natural resources, waterways and trans-

portation routes in further developing and maintaining the strength of our country. Employ charts, tables and maps to help make the concepts clear.
3. Develop with pupils: (a) a study of the diversification of the resources of the United States, and (b) the story of industrial and technological progress from colonial times.
4. Develop an understanding of democratic citizenship by reading stories about the lives of famous Americans, including the native-born and naturalized, who have contributed to our democratic way of life.
5. Have students plan a class or assembly program depicting certain important phases of the American heritage. Follow with a discussion of the responsibilities of pupils to preserve this heritage.
6. Prepare a collection of posters and pamphlets on civil defense, or suggest that pupils make their own posters for the bulletin board or display case.
7. Assign pupils topics on how eighth grade students can participate in civil defense.

World Geography and Economic Citizenship (Grade 9)

The economic world course for the 9th year should provide the foundation for understanding the geographic and economic bases upon which the various phases of our system of defense rest. At this grade level, pupils can begin to comprehend the civil defense program in its broader, overall aspects.

In the study of the shrinking world, with all the implications of speed in travel and communications, teachers should make students aware not only of the dangers facing our nation, but also of the serious responsibilities of the people for the defense of our country.

Suggested activities are:

1. Using a physical map of the world, pupils might study how natural physical features aid or restrict the exchange of peoples, ideas and goods.
2. Describe, using illustrations, the effects of climate and weather conditions on land, sea and air transportation.
3. Ascertain the air distances and routes between capital cities of the major countries of the world. Prepare a table of this information. Discuss the implications with regard to trade and defense.
4. Illustrate how geographic factors have influenced modern military operations and strategy.

5. Plan a class discussion on the influence that the airplane has exerted on nations, people of the world and your own community. Use appropriate visual aids. Plan a similar discussion on the influence and use of oil.
6. Use population distribution concepts to illustrate the problems involved in the evacuation of the homeless and injured in the civil defense program.
7. Locate on a world map the large sources of important materials used in our defense program. Devise a chart showing the kind and amount of these materials which we import and their principal sources.
8. Ask pupils to trace on large wall maps the important trade routes of the world. Follow with a discussion of how the control of these routes affects the local community and is important to defense.
9. Have pupils locate the areas in the Americas where there are concentrations of industries vital to defense. Include a study of the transportation facilities by which raw materials and finished products are carried to and from these areas.
10. Compare the cost of other government services with the cost of the civil defense program.

World History (One Year)

Although other courses in the citizenship education sequence bear more directly upon the subject of civil defense, the course in world history serves as a background for understanding the world and also the problems of our nation's leadership, and perhaps of our survival. The following are possible activities and topics:

1. Students may be interested to read and report on examples in history of civilians who rallied to the support of their nation in periods of crisis. The Dutch against Philip II, Spanish guerrillas resisting Napoleon, Norwegians defying the Nazi invaders, and Londoners withstanding the blitzes during World War II are suitable examples.
2. Make a class chart of the important wars in history, showing their backgrounds. Discuss ways in which advance preparations might have averted some of these wars or helped mitigate their effects.
3. Identify ten of the largest cities in Europe as of 1939. Show to what extent they were damaged during World War II. Find out, as far as possible, what restorations have been made to date.

4. Invite foreign-born residents of the community or citizens who have lived abroad, in order to help the pupils better understand peoples of other lands.
5. Have the class conduct forums or round-table conferences on such questions as: Does possessing information about other peoples and nations, both friendly and unfriendly, have any defense value for us? To what extent is warfare becoming more destructive than it was in previous periods of world history?

American History

An essential objective of the study of American history is the development of responsible citizenship. Girls and boys in senior high school American history courses during the crisis years of the immediate future should have teaming experiences to help them understand why their active participation in and support of the civil defense program are an inherent part of that citizenship. They should understand our priceless heritage of freedom, and learn by precept and practice how to use their time and talents to safeguard that heritage. Suggested activities and topics are:

1. Prepare, as a class project, a list of the factors, tangible and intangible, that we find in American history which encourage a high morale in facing a national emergency.
2. Develop a class chart comparing the amount of money appropriated for defense purposes in the national budget for the fiscal years 1917–1918, 1943–1944, and the current fiscal year.
3. Students may be encouraged to write articles for the school or community newspaper explaining such needs as those for blood donors, auxiliary firemen and police, aircraft warning ground observers and filter center personnel, and other civil defense volunteers.
4. Include in the study of crime: (a) the role of the police in protecting lives and property and in controlling panic during an emergency, and (b) the control of crime in evacuation and reception areas.
5. Study other problems that may arise in an evacuation or reception area, such as: (a) preparation of facilities, (b) living under hardships, and (c) social distress.
6. Follow up the showing of a film such as "One World or None" with a written appraisal by members of the class and a class discussion.
7. Arrange with the local radio station to present regularly, if possible, a student-prepared forum, quiz program or play on specific aspects of civil defense.

8. Conduct panel or round-table discussions on several of the following questions:
   a. Will a policy of containment of communism strengthen our defense program?
   b. Why is there need for cooperation among nations opposing communism?
   c. What are the important influences that the airplane has had or may have on relationships among peoples of the world?
   d. Is the atomic bomb an overrated weapon? Have pupils read widely from books and pamphlets from authoritative sources so that expressions of various opinions can be documented?
   e. How can education become an important factor in the defense of our communities and nation?
   f. What can a responsible citizen do practically for civil defense?
   g. What is the responsibility of workers for maintaining production by remaining at their machines and desks in times of national danger?
   h. What should be the dividing line between civil and military authority in such matters as control of strategic resources, production of atomic materials, air patrols, and other defense measures?
   i. Can a strong civil defense program act as a deterrent against war?

Art

Art can be used in many ways to help young people release tension and fear. It is also a useful medium in helping to establish desirable attitudes toward the community's defense program. Pupils might:

1. Design and make posters that call attention to the danger of atomic warfare; the need for preventing panic; what to do and what not to do in an emergency; and the role of the individual citizen in civil defense.

2. Illustrate articles on various aspects of civil defense for the school newspaper and other school publications.

3. Paint signs for their local civil defense organization, and prepare maps showing the location of shelter areas, the American National Red Cross center, and various civil defense installations.

4. Prepare civil defense reminders such as cartoons, illustrated booklets and leaflets, placards, and items for newspapers.

5. Provide art materials for use in shelters and assist the younger children in using them.

Business Education

The various courses in business education provide many opportunities for students to obtain firsthand information about civil defense, and to make a very real contribution to school and community civil defense programs. The following are some of the activities that may be utilized in this manner:

Introduction to Business

1. Representatives of the local civil defense office could be invited to speak to the Introduction to Business class on various phases of the civil defense program. This class might then undertake the task of setting up a plan for coordinating the civil defense activities of all business education courses.
2. Each member of the class might prepare a short speech on the part that the school plays in the civil defense program. After practicing before their classmates, these young people could be asked to give their talks before other classes and local service clubs.

Business Law

1. A representative of one of the local insurance companies might be invited to talk to the business law class on the insurance situations involved in an atomic bomb attack.
2. A discussion of the legal implications of insurance coverage in the case of loss of life or property during and after atomic bomb attack might be conducted.

Business Arithmetic

The cost of setting up a local civil defense organization might be studied as a class project in business arithmetic.

Business Management

1. As a project in business management, the class could discuss the precautions which should be taken by a family in order to protect adequately or salvage their personal and real property in the event of an atomic bomb attack.
2. This class might also undertake the management of school civil defense funds used for the purchase of mimeograph supplies, poster materials, etc.

Secretarial Practice

1. The civil defense materials available in the local community can be dictated to the Secretarial Practice class in "office-style dictation." The oral transcription of these materials might prove instructive and valuable.
2. The Secretarial Practice class could prepare an emergency civil defense bulletin for the local civil defense committee to be used in preparing for an atomic bomb practice alert that is to take place. The cutting of the stencil and the duplicating could be so timed as to approximate actual emergency conditions.
3. A centralized file of all civil defense materials for the entire school could be set up by the Secretarial Practice class. These materials should include newspaper clippings, pamphlets, bulletins, posters, etc.
4. The Secretarial Practice class could help handle the clerical details as might be necessary in the operations of the local civil defense office. This would include the preparation of form letters, reports and notices, and the maintenance of records, files, and rosters.

Cooperative Classes

Each part-time worker might prepare a report to be given orally to the cooperative class as a whole, covering the various regulations with regard to civil defense in effect at the business establishment where he works.

Typing

1. As a typewriting project, each member of the class might prepare a master copy to be run off on a fluid ditto or similar type machine. This copy might be a memorandum to all civil defense volunteers, or some like activity. The subject matter for the memorandum should be something needed by the local civil defense committee.
2. Assist the secretarial practice class in handling the typing necessary to the operation of the local civil defense office.

Health Education and Physical Education

It is the purpose of these programs to help our high school youth become physically, mentally and emotionally strong. All aspects of health education and physical education are thus very important to the civil defense effort, and should be given increasing emphasis in these troubled times.

Certain specific areas of content within these programs provide opportunities for teachers to give instruction that relates directly to preparedness for meeting emergencies. The following suggestions are not intended to be all inclusive, but are presented in the hope that they will help teachers to find many possible ways of getting their students ready to function intelligently and effectively should an emergency arise.

Mental Health

Mental health instruction can be utilized to discuss with students the nature and causes of panic, bringing out ways in which the individual can help himself and others to hold normal fears within bounds. It is desirable to stress the need for being ready, for knowing what to do in emergencies, as a means of promoting security and minimizing fear. It can be pointed out to young people that other areas of health education and physical education, as well as many of the remaining subjects in the curriculum, contain much that will help them to learn a great deal about what should be done in times of unusual stress.

Health Protection

This area of instruction calls for considerable emphasis to be given to ways of protecting one's own and others' health. Many topics, such as the sanitary use of emergency drinking and toilet facilities, safe storage and preservation of essential foodstuffs, and the planning and utilization of shelters to minimize personal injuries from bomb blasts—all these and many more, are important learnings in civil defense.

Safety

This is an area of instruction which lends itself to the teaching of safe conduct during emergency drills, to home and school preparation for fighting small fires, to rules for staying off highways essential for defense use, and to many other learnings which will help young people take the necessary precautions to minimize injuries caused by enemy attack.

First Aid

An area of instruction which is very important at all times, first aid should receive a great deal of attention. Young people in junior and senior high schools are mature enough to benefit from a detailed presentation of all phases of first aid. First aid measures likely to be needed by blast victims should be carefully taught, assuring that our young people will be ready to help themselves and others who are injured during any emergencies.

Physical Education

A full and rich program in physical education is offered in every high school in the State. Physical education's aim is to bring about proper conditioning and intelligent use of the body. When daily tasks can be performed with a minimum expenditure of energy, through skillful use of the body, there will exist a reserve of physical power to be used in time of emergency. In this way the physical education program can contribute to our total defense effort.

School Camping

This is an area in which it is possible to give young people actual experiences in living under more or less primitive conditions such as might exist in an emergency. In this program, it would be a very good idea to give junior and senior high school students opportunities to share responsibility for the care and instruction of younger children.

Shelter Activities

Secondary school students can be of great assistance to young school children when they are taken into shelters for protection—by providing interesting activities, rhythms and games. Also, there are a number of activities in which these young people themselves might engage to prevent or relieve emotional tensions during periods of shelter confinement. Some of these are:

Self-testing Activities
Full Top Spin
Upspring
Bells or Clicks
Swagger Walk
Corkscrew
Hand-Wrestle
Egg-Sit
Chinese Get-Up

Active Games
Clap In
Hotball
Pencil Relay

Chair Relay
Numbers Up
Necktie Relay

Quiet Games
Third Degree
Buzz and Fizz-Buzz
Word Lighting
The Ship's Record
Rigamarole
Doublets
Shakespearian Romance
Zoo

Home Economics

The program in home economics for junior and senior high school youth can do much to encourage students to make home preparations for emergencies. Teachers of these courses might help their classes to:

1. Consider some of the things that can be done to prepare homes for emergencies, such as keeping a several day supply of food on hand; having emergency cooking facilities available; making sure that first aid kits are ready; and keeping on hand a supply of essential clothing in an easily accessible place.
2. Consider some of the difficulties that may arise in times of stress and the ways to deal with them. For example, what could be done about separation of family members, change in housing facilities, quartering of refugees, and many other problems.
3. Develop an understanding of the needs of children and old people, and determine ways to meet their special needs in emergencies.
4. Develop the ability to prepare simple, nutritious meals from emergency supplies and with emergency cooking facilities.
5. Plan ways of working with and understanding younger children so that they may assist in their care during emergencies. Wherever possible, high school students should be provided with actual experiences in working with children in shelters, playgrounds, camps, cafeterias, nurseries and kindergartens.

Industrial Arts

It is possible for students in industrial arts classes to contribute directly to the civil defense programs of their schools and communities.

They can, for instance:

1. Produce metal identification tags for school pupils. Use German silver and metal letter stamps.
2. Make arm bands for the identification of guards, members of safety patrol units, etc. Use a textile center for fabrication and silk screen work for lettering.
3. Prepare plans and models of air raid shelters. Draw plans for different types of shelters for various kinds of buildings and homes, including the pupils' individual homes, and construct scaled models of shelters for the homes.
4. Build emergency equipment, such as litters, crutches, splints, and first aid boxes, and also toys and games for use during stays in the shelters.
5. Devise teaching aids for civil defense, such as posters, microfilms, leaflets, and models.
6. Organize an engineering squad to:
    a. Familiarize pupils with the building details, including the utilities.
    b. Shut off the utilities and support other protective services.
    c. Make emergency repairs.
    d. Restore damaged utilities.
    e. Perform light rescue work.

Science

The science program in the junior and senior high school has a key role in training pupils for civil defense. The work of the science teacher falls into several well-defined categories. These are:

1. Helping pupils develop a scientific attitude toward problem solving in all areas of living.
2. Imparting information to all about scientific developments behind the methods of modern warfare.
3. Developing proper attitudes in all pupils based on knowledge of facts rather than fear of the unknown.
4. Encouraging the most capable pupils to enter careers in science, medicine and engineering.
5. Building abilities and skills in all for taking care of themselves and others in emergencies.
6. Screening current news reports to distinguish facts from propaganda.

7. Emphasizing the need for a well-organized and well-trained civil defense organization.

In the specific areas of the science curriculum, teachers can select suitable activities for developing understandings related to civil defense that are appropriate for each course. In many cases the science teacher will have to clear up misconceptions which pupils have acquired about atomic energy and related problems through reading popular periodicals and from the radio, television and movies. Activities such as the following might be included in the various courses:

General Science in the Junior High School

1. Handbooks 1 and 2 are sources of many activities related to personal and public health, including emergency needs.
2. Handbook 3 (1954) will contain numerous activities on first aid and an area "Living with the Atom" consisting of activities basic to understandings of atomic energy.
3. The 89 activities fundamental to living in the atomic age found in "Living with the Atom."
4. Pupils should be encouraged to select projects related to civil defense for local and area science congresses and fairs.

Biology

1. Arrange for a member of a civil defense medical team to discuss the effects of radioactive materials on the human body.
2. Investigate and discuss the effects of an atom bomb attack on the food and water supplies and distribution of a community.
3. Relate some of the learnings on the functions of the human body to first aid practices and procedures.
4. Investigate the uses of radioisotopes in medicine and research.

Earth Science

1. Investigate the methods used in searching for minerals essential for national defense.
2. Collect information pointing up the need for conservation of natural resources, such as minerals and petroleum, in time of peace as well as during national emergencies.

Chemistry

1. Collect factual information on the A-bomb and the H-bomb basic to understanding the principles of fission and fusion.
2. Use the school's Geiger counter for detecting beta and gamma radiation from various sources.
3. Study the effects of various materials for shielding against radiation.
4. Discuss the relative penetrating power of alpha, beta and gamma radiation to provide information for the protection of individuals against these and other radiations.

Physics

1. Investigate the various methods for detecting radiation. Pupils can become familiar with and learn to operate various detecting devices.
2. Arrange for a member of the local civil defense organization's Radiological Service to demonstrate various detection instruments used in radioactive areas. (Some teachers have been trained for this work and pupils in advanced classes can become adept in using the instruments).
3. Collect information on the energy released by A-bombs and H-bombs, and compare with conventional explosives familiar to the pupils.
4. Investigate the use of radar in civil defense and its limitations which make it necessary to establish and man ground observation posts.
5. Include in the study of new developments in airplanes, rockets and guided missiles a discussion of the problems that these bring to civil defense personnel.

4

Safeguarding Democracy

A state of mobilization and the threat of war create varied and unique problems involving the safety of children. To meet these problems, educators must have all the information necessary for sound planning and effective administration of their school civil defense programs.—Federal Civil Defense Administration[1]

When President Harry S. Truman delivered an address at Fordham University in 1946, his message was unequivocal: America's essential task was educating its children in order to protect the nation's future. If civilization is to survive, Americans of all ages needed to "cultivate the science of human relationships—the ability of all peoples, of all kinds, to live and work together, in the same world, at peace," Truman said. "When we have learned these things, we shall be able to prove that Hiroshima was not the end of civilization but the beginning of a new and better world." Truman then called upon America's school-age children to learn the meaning of democracy because the fate of the world depended upon them. "I know that education will meet that challenge," Truman concluded. "If our civilization is to survive, it must meet it."[2]

Richard Noyes, chemistry professor at the California Institute of Technology, echoed Truman's emphasis on the significance of the "science of human relationships." In a speech before the Southern California Science Teachers Association the same year, Noyes urged attendees to encourage their students to enroll in both physical and social sciences. "[T]he present lack of any branch of knowledge that can rightly be called a science of human behavior and relationships," he told the audience, "should not be used to scare able young people away but should be thrown in their faces as a challenge. I ask you as teachers of the generation that must live

**UNITED STATES
CIVIL DEFENSE**

CIVIL DEFENSE
IN SCHOOLS

TM-16-1

FEDERAL CIVIL DEFENSE ADMINISTRATION

The Federal Civil Defense Administration produced some 475 million pieces of civil defense materials during the 1950s, with many of its publications aimed at parents, children, and schools.

with the atom bomb to encourage your best students to face this challenge."³

As historians Willis Rudy, Andrew Hartman, and others have chronicled, the early postwar years could be called "the Age of Adjustment" for America's schools, which underwent a paradigm shift from a child-centered approach to one of life adjustment based on the tenet of education for a democratic social order.⁴ The creation by the Office of Education of the Commission on Life-Adjustment Education for Youth, which existed from 1947 until 1954, reflected the federal government's commitment to teaching skills considered necessary to adjust to the new world. Adolph Unruh, writing in *Progressive Education*, published by the American Education Federation, defined this form of education as equipping youth to live democratically and to benefit society as "home-members, workers, and citizens."⁵

Repeatedly from the late 1940s through the 1960s, teachers were instructed to emphasize the democratic way of living: an idealized notion, perhaps, of a nation based on justice and understanding, racial and ethnic harmony, and a spirit of domestic and international tolerance and friendliness. Incumbent in this effort was a defense of educational freedom to these principles without outside pressure, or in some educator's view, propaganda. Chester Diettert, a high school principal from Indiana, stated this viewpoint in a 1950 issue of *School Activities*:

> The chief danger to democracy is the dual monster of propaganda and gullibility. Propaganda without gullibility of the people will not work. America, with its free system of education (academic freedom) and free education for all, has the best safeguard against propaganda of any nation in the world. We are fighting gullibility dynamically. Some nations use the gullibility of their people, the result of mis-education or lack of education, to make effective the propaganda handed out by the government. People who do not have the safeguard of education free from propaganda remain gullible to more propaganda. The result is the inevitable submerging of individual interests to the state. No happy balance can be maintained, no natural benefit or mutual advantage results. Dictators thrive on such conditions.⁶

By the mid–1950s, obstreperous critics had forced the progressive education movement into retreat, blasting school curriculums based on social, mental, and emotional adjustment rather than intellectual achievement. Historian Lawrence Cremin has suggested that the movement collapsed primarily because of excessive demands on teachers' time and ability, growing negativism and distortion about its aims, and the failure of progressive educators to keep pace with continuing transformations within American society.⁷ Such diatribes as Arthur Bestor's *Educational*

118 Atomics in the Classroom

Wastelands, Albert Lynd's *Quackery in the Public Schools*, and Mortimer Smith's *And Madly Teach* brought into question many of the principles deemed crucial by teachers, educators, and even government officials at the dawning of the atomic age.[8]

A closer examination of educational articles during this period, particularly those endorsing the fundamental need to prepare youth for the new world, finds that concomitant with themes of democracy was the ever-present threat of atomic apocalypse. The educational community's

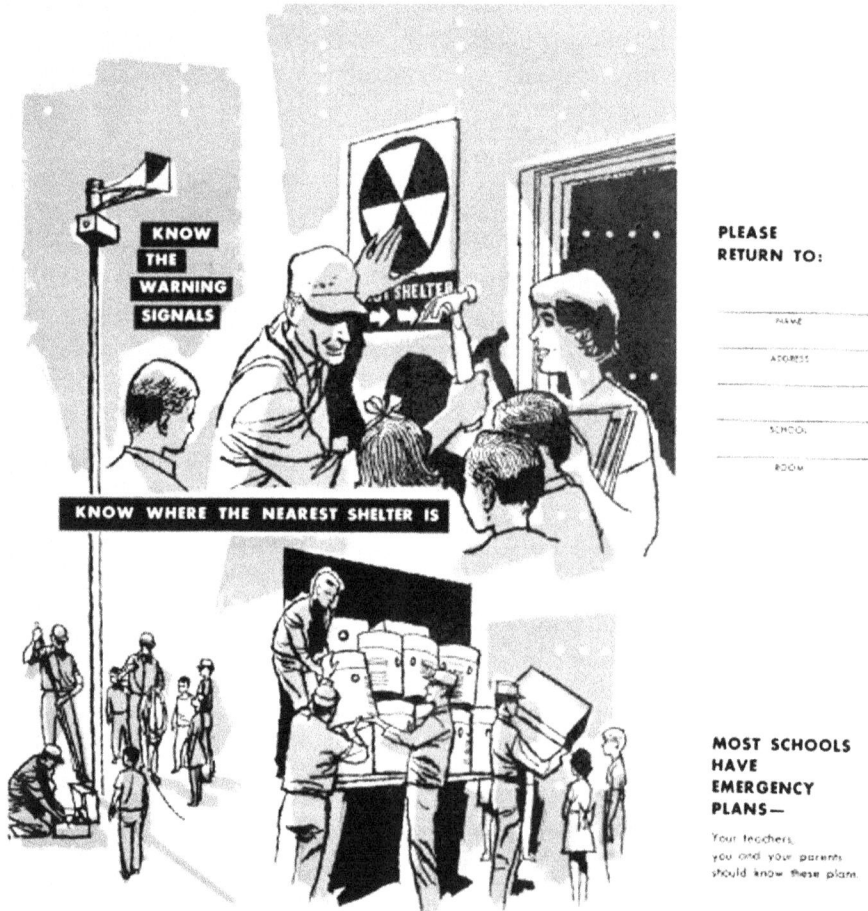

Book covers such as this one represented another reminder for students that they needed to be prepared at all times for an atomic attack.

emphasis on democratic principles throughout these years had the underlying theme of world survival in the atomic age. Regardless of the stated theme, the central issue remained peace or war, which in the atomic age meant life or death. If the United States were unable to contain communism (the totalitarian nemesis and threat to people everywhere), in other words, the country would surely perish along with the rest of the world. Following Hiroshima, life-adjustment education became an important adjunct to the more crucial requisite of the atomic age: survival of democracy and civilization itself. And teachers repeatedly underscored coping and survival skills, as well as the importance of cooperation and the submission to the group.[9]

As Samuel Capen, chancellor of the University of Buffalo, explained to an audience in June 1947, the only remaining form of totalitarianism in the world was communism. Furthermore, he said, the nation would survive only if everyone recognized and understood the nature of this inimical force, and knew how to confront and to defeat it. Whether the United States wanted to or not, fate had made it the leader "in the worldwide struggle to restore the ideal of the truth to the judgment seat before which the causes of men and of nations are brought to trial." People would only be able to understand this force, however, if the barrage of propaganda (or "red-baiting") against the educational community abated. That was the message of Curtis MacDougall of the Medill School of Journalism at Northwestern University in his address before the 1947 National Congress of Parents and Teachers. "If we fail to do so," he said, "an Atomic Age war can mean only our annihilation. Then Adolf Hitler, through losing the battle, will have won Mein Kampf." Educators, like Capen and MacDougall, constantly reminded others about the close connection between democratic values and the inherent dangers of the atomic age.[10]

John W. Studebaker, U.S. Commissioner of Education, wrote and spoke often on the challenge of education in protecting the world from what he termed "atomic suicide" by promoting democracy as the best means of safeguarding the future. In a 1948 lecture, he framed the current international situation as a world in collapse behind an "Iron Curtain" of communism, referring to Winston Churchill's speech two years earlier. Countries in eastern Europe had fallen victim to the avaricious Soviet Union. Italy was unstable. Totalitarianism was marching through Asia. Democracy was under attack; its own future in jeopardy. "Democracy could conceivably be destroyed," he said, "by a purblind and complacent belief that it can never really be in mortal danger from any rival ideology or form of human political association." If democracy were to survive, education must develop a better appreciation for democratic principles among the next generation, the recipients of this democracy. He set the tone for subsequent educational tracts when he called upon high school students to learn about communism in order to recognize its pernicious traits. Before adolescents could understand what they are against, they needed to understand what they were for, according to Studebaker. Simply stated, "education in a democracy must unequivocally be education *for* democracy." Yet, as Studebaker pointed out, the younger generation was maturing not only in an atomic age but an age of propaganda, which exacerbated a teacher's challenge. The war was not merely between democracy and totalitarianism (or communism); it was between the eventuality of the life or the death of civilization. As a result, merely teaching democratic facts and principles was inadequate. In Studebaker's view, educators had to create a "zealous dedication" to the defense and implementation of the democratic faith.[11] This mandate reflected the alarming results of a September 1947 Gallup Poll that found fifty-eight percent of those responding believing war was inevitable within ten years, and seventy-five percent believing war would occur within twenty-five years.[12]

In 1948, the U.S. Office of Education responded to rising concerns about another war by introducing the "Zeal for American Democracy" program, which was quickly endorsed by the National Council for the Social Studies and National Council of Chief State School Officers, among others. Twenty-two educators, meeting in Washington, D.C., in early March, passed a resolution calling upon teachers to strengthen the teaching of democracy by emphasizing loyalty, civic duty, creative spirit, respect, and the constant search for truth, with the ultimate objective of offsetting the growing fear, apathy, and pessimism among school-age children. To accomplish their objective, teachers were encouraged to use radio, magazines, newspapers and other outside materials in their classrooms, and to develop class projects and school assemblies aimed at demonstrating the threats to democracy and world peace. Books and booklets to be read by teachers and students covered such topics as democratic living at school, safeguarding civil liberties, the meaning of democracy, democratic human relations, and citizenship in a new world.

In support of the program, *School Life*, published by the Office of Education, dedicated a special issue to the theme "Zeal for American Democracy: Education to Meet the Challenge of Totalitarianism." Included was an article outlining the legal obligations of teachers to promote a devotion to and an understanding of American ideals and governmental principles. The article also featured a sampling of state requirements on school obligations. A typical example was the entry under Maryland, which stated, "The love of liberty and democracy, signified in the devotion of all true and patriotic Americans to the flag and to their country, shall be instilled in the hearts and minds of the youth of America."[13]

Charles Peters, a visiting professor of education at the University of Miami, working in conjunction with the Zeal for Democracy program, compiled a book based on what he labeled "democratic, action-centered (DAC) education." Junior high and high school teachers used the DAC method in their history, social studies, civics, sociology, and psychology classes to teach the tenets of democracy along with current affairs. In Bedford, Pennsylvania, for example, a high school history class conducted panel discussions on the United Nations and its chances for success as a means of preventing an atomic war. To encourage other teachers in other parts of the country to adopt the DAC method, Peters send approximately 8,700 books free of charge to teachers nationwide.[14]

Peters's efforts paid off. Ron Davis of the Teachers College at Columbia University in New York City subsequently introduced a similar program called Citizenship Education Project, in which students participated in the Ground Observers Corp, spoke before community groups on the importance of civil defense, and helped with school disaster drills. In class, they studied the importance of good citizenship and defending the nation against communism, and were required to complete oral and written reports. Introduced initially in eight cities, the Citizen Education Project eventually expanded to include close to 1,000 schools, 2,000 teachers, and 55,000 students in thirty-seven states.[15] At Quincy (Massachusetts) High School, students studied propaganda, economic security, minority rights, crime prevention, and family stability in their Problems in Democracy course. And in the State of Kansas, the Eisenhower Foundation sponsored a citizenship program aimed at youth not attending college.[16]

The belief that teachers must appeal to the hearts and minds of youth had been addressed by others. For example, W.H. McFarland, supervisor of the Iowa Department of Public Instruction, had written in 1946 that concentrating on the emotions, or heart, of youth was a prerequisite for the conveying of the harsh realities of the atomic age.[17] In a presentation before the 1947 convention of the National Council for the Social Studies, and reprinted in the February 1948 issue of *School Life*, education commissioner John Studebaker argued similarly that teachers must "inculcate in the minds and hearts of our American youth the basic principles and the fundamental ideals of our American way of life, to create zeal for American democracy." Educators had the primary responsibility for teaching youth the understandings, the skills, the attitudes, and ideals constituting American citizenship and international cooperation for peace. Moreover, educators should teach these attributes with enthusiasm and assurance, according to Studebaker, "convinced that the understandings and consequences we develop in young people today will enable them to carry forward with unflinching determination the torch of freedom, justice, and humanity tomorrow." Again, Studebaker sustained the interdependence of safeguarding freedom and the survival of humanity with adherence to democratic ideals. As for youth, they must learn to feel democracy, to have a passion for its defense, to integrate its principles into their everyday living. In this way, democracy would survive, and so too the world.[18]

David Lilienthal, chairman of the Atomic Energy Commission, addressed the 1947 American Education Federation conference on this irrefutable connection between democracy and destruction. "The plain fact is that unless the American people as a whole become informed so that they can chart the course of their own destiny in the atomic age," he told the educators, "then democracy in its very essentials is doomed to perish, to perish not by the action of a foreign foe, but by default by our hands—by the hands of those who love it." In other words, atomic energy in the hands of people who don't fully understand its implications is a threat to the future of democracy. And unless Americans have the real facts concerning the atom, and participate in "fateful matters," democracy is lost. Teachers, Lilienthal proposed, must teach the principles of self-education, so that the younger generation would learn how to act intelligently about atomic issues. This meant that schools should integrate atomic energy education into science, English, and the humanities—the atomics curriculum. Lilienthal's concern was democracy; his fear was atomic catastrophe.[19]

"[E]ven as the world still suffers from its recent devastation," wrote the Office of Education's Kendric Marshall in 1948, "we face the fact that the peace which we had believed to be securely won is in jeopardy." Communism, defined as an "aggressive ideology," threatened the democratic way of life, particularly among the war-weary people of Eastern Europe. The world was slowly evolving into two antagonistic groups: the United States and the Soviet Union, democracy and communism. The intertextuality of democracy and atomic warfare was clearly evident in Marshall's statements: "[T]o teachers as to millions of other citizens there comes the haunting fear that upon our aching generation will be unleashed atomic bombs and guided missiles of frightful destructiveness." Five years later, another educator again framed democracy within the constant atomic threat when he wrote, "The hard fact is that we in America have become aware that atomic energy is a controlling factor in the survival of our democratic way of life and that if the freedoms we cherish are to be preserved this new force must be understood by our people."[20]

As the struggle between democracy and communism intensified, so too did the emphasis on the new generation to become good citizens. By the late 1940s, citizenship had become another significant element in educational (i.e., atomic) narratives as schools adopted rhetoric and programs designed to instill democratic values in their students. Teachers were encouraged to educate students in the three C's: citizenship, courtesy, and cooperation. During these uncertain and seemingly precarious times, cooperation was especially important. This was, after all, merely an extension of the concept of brotherhood, the requisite for survival of the species and Earth itself. Teachers also stressed the three C's in vocational courses so that non-college-bound students developed proper work habits, attitudes, and skills; for students in non-vocational courses, the three C's were oriented toward making better citizens and family members.[21]

In New York State, the acting commissioner of education, Dr. Lewis Wilson, offered a four-point program for building good citizenship. Schools should first increase their emphasis on American history, particularly on current world conditions, according to Wilson. Second, mathematics and sciences should be stressed (particularly for boys) as well as pre-induction courses for those who may eventually enter the armed services. Physical fitness programs and improved health examinations would ensure a healthier generation of citizens. And, finally, Wilson said, teachers and students alike should be well trained in all aspects of first-aid, in case of the inevitable.[22]

Writing in the March 1951 issue of *The High School Journal*, J. G. Umstattd of the University of Texas called on teachers to "add strength, sanity, and unity to our nation." This could be accomplished by extending health and physical education, intensifying the teaching of moral issues, and creating mental health programs. In continuing defense of progressive—or liberal—education, he also reminded educators that students should be trained in propaganda analysis, which would result in "less anxiety, less mass hysteria, and less aid to the enemy by character assassinations." Reflecting the words of Studebaker, Umstattd argued that understanding the myriad social organizations around the world, including communism, should be coupled with the teaching of tolerance for those of different nationalities, races, and color. "To know the habits of rattlesnakes helps one to eradicate them," he wrote. "So it is with communism. A study of that way of life is our best protection against it, a far more intelligent approach than to make it the victim of our bigotry." The underlying message was not only the inseparableness of democracy and peace, however; it was about the preparation of a generation to assume its social responsibilities and demonstrate the willingness and ability to act upon fundamental values, beliefs, and ideals within a truly democratic context:

> Democracy itself must be studied in comparison with all other modes of life, and it must be lived in school and community. The secondary school program must lead in the development of those ideals and habits of living upon which the future of peace and democracy throughout the world depends. Unless we are to violate the right of decision which is fundamental to democracy, children and youth must learn by their own free will the meaning of freedom and must practice the proper ways to use it; they must of their own volition sense the meaning of civil liberties and gain skill in habitually meeting their attendant responsibilities. They must through choice realize the value of cooperative effort and master its essential activities.[23]

That same month, *Education* dedicated its issue to the theme "Educating for Responsible Citizenship in the Atomic Age," edited by R. Will Burnett of the University of Illinois. The atomic age would become a Frankenstein monster if apathy were allowed to spread, wrote Burnett. To prevent this, teachers were instructed to stress the positive, or peaceful, aspects of atomic energy. Sumner Pike of the Atomic Energy Commission reinforced this sanguine message by examining the atom's medical, agricultural, and industrial applications. If Pike was to be believed, the peacetime promise of atomic energy had already turned into actuality. Warren Austin, U.S. ambassador to the United Nations, concurred with Pike on one point: Unlocking the power of the atom had opened up new opportunities to control the environment and improve the quality of life for everyone. In his opinion, however, the paramount issue confronting the world was not the benefits of radioisotopes but the control of atomic weaponry. "No one can know with certainty just how long the free world has in which to build up and unite its defensive strength beyond the deterrent power of American atomic supremacy," Austin wrote. "But of this I am convinced, the survival of all depends now upon rising above the terror propaganda and using every moment to build a united strength." The peace of the world depended upon the ability to control atomic development, an effort being thwarted by the Soviet Union. Again, within Austin's text of democracy vs. communism was the apocalyptic vision of atomic war. "We ... know that the Soviet leaders are following a diabolical design of world domination which they call 'liberation,'" he wrote. "And we ... know that the free world possesses potential power roughly twice that of the common enemy. Our task is to use what time we have to mobilize and coordinate that potential power. If such a mobilization does not prove to be an effective deterrent to aggression, then at least the free world by its collective preparations will be in a better position to resist and ultimately to triumph."[24]

Homer Higbee of Michigan State College disputed Pike's position, suggesting that peaceful applications of atomic energy were not being realized because of the government's secrecy and control of atomic research, resulting from international tensions with the Soviet Union.[25] Furthermore, atomic energy was not being used widely in private industry because of the high cost of building reactors and producing plutonium, and the inability to control radioactive gases. In Higbee's view, the prospects for peaceful atomic developments were overshadowed literally and figuratively by the mushroom cloud. Maintenance of the secrecy surrounding the atomic bomb—"the omnipresent watchword of our time"— had encumbered consequential peaceful development. The discovery, use, and subsequent control of atomic energy, in fact, had shaken the foundation of the country's social structure, forcing it to adapt to technological change rather than constrain technology within acceptable social parameters of advancement. "Before we can gather our bearings and plot a successful course for the constructive use of atomic energy," he indicated, "we are faced with the problem of whether or not we will use the A-Bomb again as a weapon of war." Memories of Hiroshima and Nagasaki still plagued the world, as far as accepting atomic energy as a social elixir. "Since the first military application of the atomic bomb," according to Higbee, "the whole world has taken on aspects of ants whose abode has been smashed by the heel of some careless walker. There has been evidence of that same aimless hurrying and scurrying to nowhere in our efforts to appraise the significance of atomic energy." Overcoming this, of course, was the goal of educators, who were required to learn as much about atomic energy themselves in order to teach their students. Higbee, although pessimistic about the current status of atomic energy, nonetheless agreed with Pike's central tenet that the atom did, indeed, hold promise. But this promise would never be realized without concerted efforts within the schools to overcome the escalating fear among the public, which had been left in "a penumbra of fear, speculation, and confusion."[26]

The world had learned to fear the atomic bomb, and this, in turn, had prevented it from understanding its motivating force, atomic energy. "Fear has created a desire to flee from thoughts of potential destruction," Higbee wrote. "A rather typical reaction is observed. Many are covering their eyes with their hands to blot out that which they do not wish to see." This fear had resulted from the complexity of atomic issues during a time of increasing paranoia and fear of communism, both as an internal and external threat. The "witch-hunting since the end of World War II," in Higbee's words, had contributed adversely to a better understanding of the atom and communism, and, consequently, had aggravated the degree of apathy, ignorance, and fear among Americans. Democracy was the only hope to prevent atomic destruction, while communism (and by implication, a Soviet atomic attack) was the major threat to democracy. Only educators, by disseminating accurate information about atomic energy to their students, could thus allay these fears and, in the process, fulfill their patriotic responsibilities in preparing the next generation for its role in safeguarding democracy.[27]

The U.S. Office of Education, in recognition of increasing world tensions—the escalation of this "hazard-freighted crisis"—and the subsequent necessity for preventing the disaffiliation of America's youth from their democratic responsibilities, made "Citizenship for an Atomic Age" the theme for the March 1953 issue of *School Life*. The "atomizing" of citizenship encompassed knowledge and understanding of the present world situation, defined as a "crisis in fear" resulting from three pernicious factors: The increasing control over nature had occurred without a comparable control over the actions of society; democracy had evolved without adequate preparations of the populace for self-rule; and the atomic bomb had made the world interdependent without "either a pervasive toleration of differences or an adequate system for dealing peaceably with common affairs." To counteract these factors, teachers must develop a consensus for coping in the current time of stress and crisis.[28]

According to Ryland Crary, who authored *Operation Atomic Vision*, the postwar world had created a paradox of hope and fear, of affirmation and negation. Science and morality, moreover, continued to exacerbate the social tensions. "It remains a matter of choice—moral choice—whether man is to elect to emancipate or exterminate himself through this science." The next generation faced profound obligations on citizenship within the atomic age. Democracy was on trial with its mortal enemy, Crary said, and, as a result, citizens' obligations extended from problems within the community to "the arena of responsible judgments on world affairs." The burden, of course, fell on educators, who were charged with the responsibility of building civic responsibility. This, as Studebaker, Umstattd, and others also posited, required an appreciation for the whys and wherefores of facts and so-called truths. Equally important, teachers must counteract fear within their students as they developed emotional security and they must develop skills for handling any potential emergency. More important, "education must pursue its great constants, its quest for the meaning of truth and beauty and the good life."[29]

What becomes readily apparent when surveying educational journals during this period is that teachers and administrators received a continual barrage of cogent arguments about their democratic responsibilities. A 1952 National Education Association report, for example, proclaimed that world war was not inevitable if western democracies built strength of arms and strength of spirit. "Our national security," the report read, "requires that we win the worldwide battles of ideas with weapons transcending those we have used in the last six years. To forge these weapons is the urgent task of education."[30] J. Clyde Johnson, of the North Carolina College of Agriculture and Engineering, even outlined a twelve-point checklist of the essential skills and attitudes of an effective citizenship in a democracy—skills and attitudes to be inculcated into school-age children:

1. Self-reliance.
2. An attitude that doesn't tolerate discrimination.
3. Regard for the rights and needs of others.
4. Respect for and recognition of those in positions of authority.
5. A willingness and ability to work with others.
6. An appreciation of democracy's historical heritage.
7. An understanding of the meaning of democracy.
8. Recognition of the individual's important role and responsibilities in a democracy.
9. An ability to adhere to majority opinion with due respect for minority rights.
10. Patriotism.
11. An achievement of reasonable conformity to group demands.
12. A willingness to work for evolutionary change, not revolutionary change.[31]

Students must learn these qualities of effective citizenship, not merely memorize them, Johnson warned. His position was clear: True democracy required that ideologies and concepts be understood, and to this end, teachers should incorporate all their necessary talents and materials to, again, reach the hearts and the minds of youth.

Johnson's program was supplemented in 1955 by Arnold Perry, dean of the School of Education at the University of North Carolina. Writing in *The High School Journal*, Perry outlined a similar list of educational objectives that included vocational efficiency, democratic citizenship, effective group participation, tolerant attitude, recognition of propaganda, and an appreciation of freedom. Despite criticism that teaching democracy was a form of indoctrination, Perry argued, children needed to grasp the principles and values of democracy so they would be able to defend the nation against all enemies. "The rising generation," he wrote, "must understand that our educational institutions are instruments of social change.... There are people who honestly believe that a thorough teaching of the Declaration of Independence, the Constitution, and the history of the United States is all that is necessary for development of youth for effective citizenship in a democracy. As essential as these activities are, they do not necessarily produce individuals capable of assuming the privileges and responsibilities of democratic citizenship. The only way individuals may learn to live democratically is by experiencing democracy. If we are to teach democratic ideals, then these ideals must form the basis of classroom procedure. Only when the school becomes in effect an emerging replica of a democratic society in action will it fulfill its obligation to the people it serves."[32]

Educators throughout this period heard this message repeated many times. They were the paragons of the proper knowledge, attitude, and behavior required to adjust and to survive in the atomic age. Their job, their responsibility, their obligation was to render their students for the challenge ahead. Among the attributes deemed necessary among elementary and secondary students throughout these years were preparing to defend the country, adjusting to their prescribed social (i.e., gender) roles, understanding the realities of the atomic age, embracing the democratic process, and becoming good citizens. And, as John W. Studebaker and others had argued, educators were to appeal to the hearts and minds of youth. This was again reinforced in 1952 by Howard Hightower, assistant superintendent of schools in Mattoon, Illinois, who wrote, "I would like to submit that peace begins in the hearts of children, and that these hearts can be made happy and secure everywhere. Education in the school, the home, and the church has the opportunity and the obligation to help build men and women with hearts and minds that will establish the foundations of world peace."[33]

Controlling fear in the hearts of children was even more critical, according to Earl James McGrath, U.S. Commissioner of Education, who called fear, want, and insecurity "the breeding ground of communism." In a collection of addresses published in 1951, titled *Education: The Wellspring of Democracy*, McGrath argued further that teachers should take loyalty oaths to disarm their critics, and that communism should be taught in the schools so that young people fully understood the nature of the country's adversary. Echoing his predecessor, John W. Studebaker, McGrath said, "Love of and devotion to the best in American life will grow only in the minds and hearts of those who know full well what the alternatives are, and who, in their own lives, have entered fully into the experience of democracy." Youth must not only understand democracy, however; they must maintain a keen interest in it on a local, national, and international level. This was the most urgent responsibility of the schools: cultivating "proper attitudes of citizenship in a democratic society, attitudes which help the student define the role government plays in the good society, and the role the citizen plays in good government." Thus, fear, resulting from uncertainties in the atomic age, became linked to the subversive techniques of communism: to be afraid was to be un–American.[34]

"Creative expression" was the phrase used by Louis Kaplan of the Oregon College of Education. He wrote in the *Journal of Education* that schools must foster creativity, ingenuity, and originality, beginning in the early grades with finger-painting and continuing on through music, art, and other problem-solving subjects. For Kaplan, the timeworn techniques of education no longer sufficed because the atomic age had obliterated the world upon which these techniques were based. Fundamental skills— the three R's—were not applicable to a rapidly changing society where the future remained uncertain and ever-changing. According to Kaplan, schools must enable children to develop the only quality that would serve them in the precarious world: "the ability to create new patterns of behavior in a world where only those who can adapt will survive." Wrote Kaplan,

> If anything is clear, then, it is the inescapable fact that, if adults are to acquire the ability to fashion new ways of life and living in an unpredictable future world, then the schools must now make a deliberate effort to foster creativity, ingenuity and originality in children and not as a haphazard or incidental program. It must be one which begins with the child who first enters school, and sweeps him along into expanding areas of challenging, stimulating experiences until he emerges from school as a self-adequate individual with a capacity for meeting real problems and solving them intelligently.[35]

Schools, he argued, should institute a planned program where aesthetic experiences merged with intellectual experiences without sacrificing the former. This coupling would strengthen the teaching of those individual qualities necessary in a complex society, "where only by finding new solutions to heretofore insoluble problems can the human race preserve itself."[36]

For teachers throughout the nation, nothing had changed dramatically since the fall of 1945, just weeks following the atomic bombings of Hiroshima and Nagasaki, in terms of instilling in their students the knowledge and skills critical for coping in the atomic age. The intensity level had changed, though, beginning with the Soviet Union's testing of an atomic bomb in 1949, and escalating significantly with the development of the hydrogen bomb (1952 for the U.S.; 1953 for the Soviet Union). These developments were then followed by President Dwight D. Eisenhower's massive retaliation—or mutually assured destruction—defense policy in 1954, with both counties armed with long-range bombers and nuclear missiles capable of hitting targeted cities, and his National Plan for Civil Defense and Defense Mobilization in 1958, which outlined the government's plan to ensure the nation not only survives a nuclear war but wins it.

This escalation only served to apply more pressure on teachers to prepare their students for the future while attempting to mitigate their fear of nuclear annihilation. As teachers had heard more than a decade earlier, they needed to instruct their students in survival behavior while, at the same time, maintaining calmness and confidence. Elementary students needed to understand what to do in an emergency, as well as why they needed to demonstrate proper behavior during both shelter drills and evacuation drills. High school students, in turn, should be well versed in civil defense protocols and, thus, assume more responsibility, such as messengers and caretakers for younger children. Yet these instructions in civil defense and survival had to be integrated within the overriding mission in the Cold War of teaching the principles of good citizenship, patriotism, and democracy, deemed crucial to safeguarding the country and defeating communism.

Educating for Survival

The Federal Civil Defense Administration, which began in January 1951, published more the 475 million pieces of literature during the 1950s, most of which was aimed at parents and their children in the form of information about atomic and hydrogen bombs and how to survive a nuclear attack. In 1957, the Civil Defense Education Project, under the aegis of the FDIC, prepared a handbook with suggestions for incorporating civil defense education into America's elementary and secondary schools—many of which had been

implemented for years as a result of state education guidelines and educators' own initiatives. The handbook, titled Education for National Survival: A Handbook on Civil Defense for Schools, *was published by U.S. Department of Health, Education, and Welfare. Following is an excerpt.*

Civil defense education should be part of the experience of every school age person. It prepares the student to survive physical disaster and enables him, as a future citizen, to protect himself and others, serve his community, and help strengthen the Nation in time of emergency.

Instruction in civil defense cannot be a packaged program. It is not something to be taught for a few days or weeks and then laid aside. Rather, it must be appropriately included at many points in the total curriculum where its application and utilization are compatible with ongoing classroom activities.

Many areas of the curriculum can be modified to include civil defense education. Science classes provide opportunities to teach understandings of the technology of modern warfare and natural disasters. Other teaching opportunities are to be found in the broad fields of the social studies, health, safety, and physical education. The importance of extra-class activities and their usefulness in civil defense education should not be overlooked.

Civil Defense Education in the Elementary School

The elementary school curriculum is designed primarily to train children in the habits of orderly thinking and help them to become mentally alert, observant, and emotionally balanced. What the child experiences during these early, impressionable school years usually remains in his consciousness through life. Since it is apparent that civil defense is now a permanent national institution, the elementary school is where its educational roots should be planted.

Civil defense education can be a part of instruction in the upper elementary schools, especially in citizenship, science, health, and physical education.

Citizenship Education

Citizenship education in the elementary school is focused on the child's own environment—home, school, and community. Here civil defense concepts can be introduced by teaching the ways in which the home, school, and community provide for the essential needs of life. Study areas can be expanded to include community health, transportation, community helpers (policemen and firemen), and civic contributions of local industries.

In addition, pupils study the history and geography of our country and other areas of the world. Here civil defense can be made part of frequent discussions and classroom activities. For example, in connection with the study of United States history and geography, teachers can emphasize: (1) Principal target areas and plans for their defense; (2) significance of the national resources in national defense; (3) major transportation routes and communications systems and their value in civil defense; (4) important food producing areas, including plans for emergency production and distribution of food; (5) effects of nuclear attack or major natural disaster upon our large metropolitan and rural areas.

Pupils studying current events can discuss publications clippings dealing with civil defense, our national defense program, and our relations with our allies.

Health and Physical Education

Most schools devote considerable time to health and physical education. Areas such as health, safety, and first aid can involve civil defense concepts and principles. Pupils can discuss sanitation as it applies to emergency use of food and water, safeguards against infection and various ways of maintaining health during disasters. The safety program can be used to teach shelter and evacuation drills, fire prevention, roles of fire and police departments in civil defense, and precautions against personal injury under emergency conditions. First-aid instruction may include information suitable to the level of pupils.

Civil Defense Education in the Secondary School

Many of the concepts outlined for civil defense education in the elementary school can be included in the secondary school program. Secondary school students are more mature and their courses are more highly sophisticated. Thus they can learn more about civil defense, and some students can participate actively in both school and community programs.

Certain understandings of the problems of the nuclear age, techniques for survival, and the development of essential citizenship qualities are basic to a school curriculum designed to meet present-day needs.

These understandings and personal development requirements include:

1. *An understanding of scientific and political developments leading to present world conditions.*

Educating for Survival 134

Contrast American democracy with other ideologies or ways of life. Americans believe that the individual is of supreme importance and that the State exists only to serve him. Our cultural heritage may be contrasted with that of totalitarian countries.

Study the pledge of allegiance to the American flag, the preamble to the Constitution, and other outstanding historical documents. Show how underprivileged or oppressed citizens or groups are susceptible to community ideologies.

2. *An understanding of the technical and scientific aspects of nuclear weapons, biological warfare, chemical warfare, and conditions which result in natural disasters.*

Study the development of atomic energy; its application as atomic and hydrogen weapons, weapons effects—blast, heat, and radiation; peacetime applications of atomic energy; international control problems; types of attack and potential dangers in biological and chemical warfare. Define natural disasters; outline conditions that give rise to them; and discuss the effects of hurricanes, floods, and tornadoes.

3. *An understanding of the precautions necessary for survival under conditions of war-caused or natural disasters.*

Necessity for school civil defense and its operation as part of the community program. Shelter, evacuation, reception, and support as civil defense concepts. Rules for safety in natural disasters. Precautions against effects of biological and chemical warfare.

4. *An understanding of home and community responsibilities of youth in an era of world tension.*

Value of preplanning for survival in time of emergency. Preparing for shelter or evacuation, and if resident of a non-target area, preparing to receive evacuees. Local government, history of the community, and ways for improving local conditions. Volunteer assistance to community services. Dependence of home and school on sound community planning. Cooperation with authorities in time of emergency.

5. *An understanding of the interdependence of individuals, expanded to include groups and nations.*

Value of helping each other in school. Relationships of family with neighbors, fire department, police department. Human and material resources of community and nation. Modern developments in transportation, communications, and utilities services. Show how interdependence cuts across racial, religious, political, and national lines. Services of the Red Cross, civil defense, and similar agencies in time of emergency.

6. *Development of democratic leadership that encourages individuals to assume responsibility for themselves and others.*

    Leadership qualifications attained through service on youth committees in school and community. Acceptance of delegated responsibilities, such as members of school patrol, users at school programs, or school civil defense wardens.

7. *Development of healthy citizens to insure national survival.*

    Provide opportunities to acquire knowledge of, and practice in, good nutrition. Develop physical fitness programs. Promote understanding of health habits and disease prevention techniques. Increase emotional stability and confidence through student participation in civil defense planning.

Social Studies

Integration of civil defense education into the curriculum takes place most logically in the social studies field. It can aid in attaining many of the accepted social studies objectives. Integration, however, may require increased emphasis on some units and addition of new content, materials, and methods. Teachers can reexamine their courses to determine where civil defense might best emerge.

Some aspects of civil defense in the social studies are:

1. *Citizenship education*—Opportunities and responsibilities for participating in civil defense.
2. *Local community*—Study of resources available for self-protection and mutual assistance.
3. *Expanding community*—Interrelatedness of target, reception, and support area communities.
4. *Federal Government*—Relationships between local, State, and Federal Government. Civil defense relationships at local, State, regional, and Federal levels.
5. *World affairs*—American foreign policy, international relations, the United Nations, control of atomic energy, world resources, and war and peace.
6. *The individual in democracy*—Civil and political rights and responsibilities.

7. *Civil defense and military preparedness*—Areas of responsibility and respective roles under peacetime and attack conditions.
8. *Problem-solving techniques*—Instituting civil defense programs as social, community problems, involving technical and human relationships.
9. *Group work*—Techniques of working with or within a group to achieve effective community civil defense programs.
10. *Psychology*—The study of mass thinking and acting in human relationships; hysteria, fear, panic, and human conduct in time of emergency, and the effect of civil defense preparedness on public opinion.

Suggested activities—Units in community living give the student a knowledge of the framework of local government and its relationship to State and Federal government. They outline the community services available to the individual and indicate ways in which the citizen can participate in government affairs. Here the teacher finds opportunity to emphasize the place of civil defense in local government.

Mutual assistance pacts and the relationships between local, State, and Federal civil defense will illustrate cooperation at various government levels. The objectives of local government services can be illustrated by the manner in which civil defense is prepared to serve in time of disaster. Participation in school civil defense activities will give the student experience in taking an active role in community service.

Activities listed below can be used in civil defense instruction:

1. List and discuss local, State, and national disasters in recent years.
2. Bring to class newspaper clippings on civil defense activities in the community or elsewhere.
3. Show films that will point up the importance of civil defense.
4. Assemble a library of civil defense informational materials.
5. Make a map of the community. Include facilities important to civil defense—schools, parks, playgrounds, telephone offices, civil defense headquarters, police stations, power plants, water systems, sewage disposal plants, fire houses, food stores, warehouses, industries, airports, railroads, evacuation routes, and reception areas.
6. Have a student committee visit the local civil defense director, police chief, fire chief, city engineer, communications center, and civil administration headquarters, to see how civil defense operates. Reports of interviews can be made to class or at assembly. If possible, have local civil defense representatives speak to class.

7. Organize projects in which student committees conduct school shelter and evacuation surveys, and form or join Ground Observer Corps teams.
8. If school observes "City Government Day," one student might spend some time in the office of civil defense and report to his social studies class on his impressions.
9. Conduct a vocabulary study of civil defense terms.
10. Study the school civil defense plan to ascertain how members of the class can become part of it.
11. Write editorials for the school paper on the responsibilities of students in civil defense.
12. Prepare a student's handbook on local government. Include information about its structure; officials, their duties, terms of office, election; civic responsibilities of citizens; registration and voting requirements; and the civil defense organization.
13. Prepare an outline of school and home civil defense activities for presentation to the parent-teacher association.
14. Show how divergent and conflicting philosophies of government have contributed to world tension. List the advantages of the individual in a democracy.
15. List the qualities of a good American citizen. In a follow-up discussion, students may emphasize activities to develop intelligent, reliable, active citizenship.
16. Explain the importance of polar geography to world affairs, using globes, air age maps, and reports.
17. Study world events since 1900 in the light of how they have led to the present world situation and show why civil defense has become a necessary part of everyday living.
18. Study propaganda techniques. How to recognize propaganda. Why it is essential for everyone to understand propaganda techniques, and how this knowledge is necessary to civil defense.
19. Incorporate school civil defense planning in the work of the student council or student government activities in the school.

Science

In science courses there are many opportunities to present the technical aspects of civil defense. Courses can incorporate data on technological developments which have led to modern methods of warfare and facts pertaining to energy released by atom and hydrogen bombs as compared to energy from other sources. Teachers can clarify misconceptions students may have regarding atomic explosions, fission and fusion, and radioactive fallout.

Fire prevention and fire fighting are suitable subjects for inclusion. They are treated in many textbooks on general science, as are units on health and first aid. Biology instructors might invite local civil defense or department of health representatives to address their classes and describe the effects of radiation on the human body and on food and water supplies following an attack. Other science activities may include the operation of radiological detection instruments, characteristics of various materials used in shielding against radiation, and hypothetical problems in estimating the relative penetrating power of alpha, beta, and gamma radiations.

Integration of civil defense instruction into science courses will give students a clearer perspective of the problems involved. Much of the technical information in chapter II can be adapted for use in science courses.

Mathematics

Mathematics is the basic tool of science, invaluable to our complex civilization. With the arrival of the atomic age, attention was centered on the study of relatively new mathematical concepts as being essential to an understanding of nuclear weapons and defense against them.

One background objective of mathematics instruction is to give students an understanding of recent technological developments and their impact on everyday living. Mathematics courses contribute to the development of practical skills, knowledge, and the ability to estimate potential results after evaluating causes.

The following activities are suggested:

1. Estimate speed, load limit, distance, and time requirements for intercontinental bombers and guided missiles to leave home base and reach their target.
2. Estimate time that would elapse before local attack, after enemy planes were detected over Alaska.
3. Determine the nature of radioactive materials through mathematical analysis.
4. Determine persistence of radioactive fallout from products of known type and quantity.
5. Measure floor space available for emergency shelters in school buildings.

6. Measure space in nearby buildings for reception and treatment of the injured and mass care.
7. Plot a graph showing the time required to evacuate persons from a target area to a reception center. Consider distances from various points in the assumed target area. Account for transportation facilities, speed of vehicles used, vehicle capacities, and number of persons to be transported.
8. Calculate amounts of food, in balanced diets, necessary to feed 100 adults, or 100 children of varying ages, for a period of 3 days, 7 days.
9. Calculate the quantities and cost of materials needed to build a reinforced concrete home shelter. Include cost of labor and excavation and so forth.
10. List the location of each fire-fighting unit in the school area, and estimate the time required for a specific unit to answer an emergency call.

Health Education

Good physical and mental health is an important factor in combating disaster. Thus civil defense is concerned with the prevention of disease, control of epidemics, maintenance of adequate sanitation, and detection and control of chemical, biological, and radiological hazards.

The school health program contains much that can be applied to civil defense. Teaching may emphasize cleanliness, practical means of preventing fatigue, home care of the sick, importance of periodic health examinations, and the correction of remediable physical defects. Health education also involves the practical application of information on the purchase, use, and preservation of food and drug products.

The following activities are suggested:

1. Demonstrate food pollution through simple fungus and bacteriological experiments. Show how and where to store foods safely under attack conditions. Demonstrate proper disposal of contaminated foods.
2. Plan an emergency 7-day food supply for a single home and for an entire school population. Plan a 3-day food supply for an evacuating family of four persons.
3. Plan an emergency diet for outdoor living. Show how to preserve and package foods. Plan an emergency food pack that can be carried by one adult.
4. Interview local medical personnel concerning emergency health facilities.

5. Demonstrate practical sanitary methods for preventing disease.
6. Visit a blood donor center. Trace the processing of blood from donor to patient. Assist in blood bank publicity programs.
7. Arrange for emergency use of the school as a hospital, first-aid station, or mass care and feeding center.

Physical Education

The physical education program is designed to promote physical and mental health, to utilize physical activities for social education, to provide opportunities for development of recreational interests and skills, and to contribute to healthful home and school living.

The physical education director is interested in the development of the individual through participation in a variety of physical activities. Types of activities chosen should depend on the needs of the students, indicated by ability, interest, maturity, sex, and physical limitations.

A well-organized physical education program contributes to the preparation of youth for civil defense by emphasizing:

1. *Physical fitness*—Physical education should lead to development of the physical vigor necessary for continued and sustained effort, and for resisting fatigue.
2. *Survival skills*—The program should provide a wide range of physical activities through which youth may learn self-protection in many different situations. Games, sports, swimming, tumbling, and certain combative sports will tend to replace fear with confidence when self-reliance is needed in an emergency.
3. *Leadership*—Opportunities range from single assignments to extensive group responsibility. A large number of short leadership opportunities for many students is preferred to a few long-continued opportunities for small groups of students.
4. *Group membership*—Planned activities for groups within classes encourage students to become competent, cooperative group members. Extraclass group activities serve the same purpose.

First Aid

In addition to being essential to everyday safety, first-aid training is vital to the care of casualties resulting from disaster. Available medical personnel and equipment could not always cope with the situation; supplementary services of persons trained in first aid are needed. First-aid training enables students to care for cuts and simple fractures until medical assistance can be obtained. Students learn to treat simple burns, identify and treat shock, and to handle injured persons.

The following activities are suggested:

1. Offer American Red Cross Standard First-Aid Course.
2. Visit a civil defense first-aid station or mobile hospital.
3. Interview medical personnel responsible for the organization and operation of civil defense first-aid teams and stations.

Home Nursing

In a major disaster, persons competent in home nursing must be available to supplement the number of regular nurses available. Students trained as nurses' aides can assist at civil defense first-aid stations and hospitals. Schools can take an active part in preparing young people for this service.

Home nursing education includes basic elements of caring for the sick and injured. Advanced students will be able to recognize major symptoms of injury or illness and to administer prescribed medication. They will learn how to protect medical supplies, assist during minor operations, and prepare patients for transfer.

The following activities are suggested:

1. Visit a hospital to observe the work of nurses' aides.
2. Interview local authorities responsible for providing hospital facilities for the care of disaster victims.
3. Practice care of a patient under simulated emergency conditions.
4. Prepare supplies for home care of the sick.
5. Prepare teen-agers to become more competent as "baby sitters" and for other child care activities.

Family Civil Defense Preparedness

Family self-sufficiency in time of emergency is essential. Students should be taught that the family is the basic unit of society *and of civil defense.* They should learn survival techniques at school and carry this knowledge into the home.

Consideration for others should be the keynote of all family functioning in time of distress, especially in reception and support areas. Family members should he made acquainted not only with emergency resources and assistance available from the community but with their individual responsibilities.

The following activities are suggested:

1. Plan for safety survey of home premises.
2. Become familiar with civil defense warning signals and *Conelrad.*
3. Study home shelter plans in school, then select safest place in the home for use as shelter.
4. Learn methods of home communication with police, fire department, civil defense, and other community services.
5. List supplies and equipment needed for home defense and designate storage locations for food and water supply, first-aid kit, battery, radio, blankets, fire extinguisher, hose, axe, ladder, shovel, saw, rope, covering for windows and doors.
6. Become familiar with evacuation routes from the city and location of the assigned reception area. Outline precautions necessary for use of the family car in evacuation. List articles that should he transported.
7. Inspect home for unsafe conditions and take steps to correct them.
8. Plan for reception of evacuees and for ways of feeding and clothing them.

Safety Education

Safety instruction in civil defense stresses precautions against the effects of atomic attack—injuries from heat, blast, and radiation. Civil defense protective measures extend to the home, the school, and throughout the community. Discussion centered on evacuation procedures should cover traffic problems and driver regulations.

The following activities are suggested:

1. Study charts of school building that show location of halls, stairways, exits, and shelter areas. Locate hazards and discuss ways of removing them.
2. Discuss the effects of a major disaster—fire, falling debris, broken utility lines, or, in the case of atomic weapons, radioactive fallout.
3. Study areas in and around school to select shelter sites that would offer the greatest degree of safety.
4. Practice protective techniques under conditions of simulated disaster.
5. Conduct a cleanup campaign to remove fire hazards.
6. Study safety measures observed by the military during convoy movement of motor vehicles.

5

The New Frontier

All in America want peace, but the world continues in a state of tension and turmoil. World domination is the goal of our enemies and ruthless aggression may well be their policy to attain it. An enormous and growing hostile nuclear force is poised and ready to crush us.—New York Committee on Fallout Protection[1]

Children ducking under classroom desks or huddling in coat closets and shielding their eyes during "flash" drills seemed rather pointless by 1960, when the Soviet Union had the capability of firing intercontinental ballistic missiles carrying multi-megaton hydrogen bombs. If anyone—adults or children—was close enough to see the flash and feel the blast of an H-bomb, there was not much sense in these drills. Yet "duck and cover" drills, shelter drills, and flash drills continued to be practiced on a regular basis in America's elementary and secondary schools well into the 1960s, even though evidence against practicing them continued to mount. According to a 1963 survey, for example, four-fifths of parents of school-age children said they would get their children at school if an attack was imminent, with thirty-two percent indicating they did not believe their children were safe at school in case of an atomic attack.[2]

Speaking to a Los Angeles PTA group that same year, Dr. Isidore Ziferstein, a California psychiatrist, told the audience that when children begin to question whether hiding under their desks can really protect them from a fifty-megaton hydrogen bomb, they also will begin to question their teachers, parents, and others who have recommended such drills as a valid means of survival. Moreover, he said, many children who dream of bombs dropping on them are actually expressing anxiety beyond what he considered tolerable psychological limits. Another speaker added that for most children the reality of nuclear war is beyond their comprehension. Instead, they exude a feeling of anxiety that takes the form of cynicism toward the future and the belief that fallout shelters will not save them.

This cynicism was not helped by a statement by the superintendent of schools in Wichita, Kansas. Writing in the February 16, 1963, issue of *Science News Letter*, Judith Viorst quoted the superintendent as saying, "The Wichita Public School system is in no position to guarantee physical protection to adults or pupils from a thermonuclear explosion or radioactive fallout.... It is therefore useless for the school system to conduct civil defense drills for an outmoded system of protection against a possible thermonuclear attack." She went on to mention a Washington, D.C., teacher who characterized her school's civil defense program as a "dead-end hoax," then shared the comment by a 14-year-old, who said rather matter-of-factly, "I guess I'm not going to grow up after all."[3]

Even though drills became a debatable topic as the 1950s progressed, however, atomics did not. Beginning in the mid–1950s and continuing into the 1960s, the Atomic Energy Commission, the National Science Foundation, and universities around the country trained high school biology and science teachers each summer on various aspects of the atomic age. In 1958, for example, Sister M. Henriella Reinders of Cathedral High School in Superior, Wisconsin, attended a workshop sponsored by the foundation to help outline a unit for biology students. Attendees came away from the workshop with three basic objectives for the unit: to get a basic understanding of the principles of radiation and its beneficial application to the field of biology, agriculture, and medicine; to become familiar with the destructive effects of radioactivity and to learn how to overcome and prevent its hazards; and to acquire a better perspective of the role of radioactivity in the lives of Americans now and in the future.[4]

In a booklet titled *Texans on the Alert*, published in 1956 by the Texas Division of Defense and Disaster Relief, schools were encouraged to strengthen their civil defense programs by incorporating civil defense in various classrooms and at all grade levels. Among the recommendations were the following:

- Teach first-graders to use their own first-aid kits.
- Encourage shop teachers to plan for shelter, safety, and improvised equipment.
- Require homemaking teachers to expand their instruction to

TEXANS ON THE ALERT
FOR CIVIL DEFENSE AND DISASTER RELIEF

ACTION PLAN FOR LOCAL GROUPS
A FAMILY PROTECTION PLAN
A PLAN FOR EMERGENCY MASS CARE

THE STATE OF TEXAS
Allan Shivers, Governor

Executive Department
Division of Defense and Disaster Relief

1956

In 1956, the Texas Division of Defense and Disaster Relief published a booklet titled *Texans on the Alert*, which encouraged schools to strengthen their civil defense programs by incorporating civil defense in various classrooms and at all grade levels, including having first-graders use their own first-aid kits.

include first aid, home nursing, Red Cross courses, seven-day food supply, food storage, improvised lighting, cooking, outdoor cookery, and camping.
- Incorporate civil defense into civics classes.
- Use posters on safety, civil defense, and survival plans created by art classes.[5]

Donald J. Fluke, director of the Duke University Summer Institute in Radiation Biology, commented in 1960, "Radioactivity is becoming a high school classroom staple. Geiger counters share bench space with microscopes, and fume hoods are bright with the yellow and magenta colors of the Atomic Age.... [T]he teacher can talk about atomic reactors, the balance between hazards and uses of radiation, and the whole spectrum of radiation biology in relation to policy and to society." With the contributions of civil defense agencies and supportive organizations, teachers received equipment kits that included a scaler-ratemeter with Geiger counters, electroscope, diffusion cloud chamber, supplies for autoradiography, and numerous smaller items. By the end of the decade, radiological kits had been distributed to 15,000 high schools and one million students had been taught about the effects of radioactive fallout.[6]

In 1961, the New York State Education Department published *Nuclear Survival: A Resource Handbook*, a guide for teachers of general science, physics, chemistry, biology, mathematics, social studies, home economics, industrial arts, physical education, and art. The handbook recommended the implementation of a civil defense awareness program to involve the entire school, including having assemblies on survival and protection topics, and encouraging students to become involved in local civil defense groups. "Participation by as many of the student body, teaching and custodial staff as possible is desired in the civil defense organization; the greater the participation, the greater the dissemination of information. When correct information is widely publicized, the panic that normally arises from a confused and ignorant population will be lessened considerably."[7]

The handbook also recommended an atomics curriculum by listing specific activities for each subject area to help students gain knowledge about and understanding of the facts and principles of nuclear radiation. In social studies, students should conduct panel discussions on such topics as why education is an important factor in the nation's defense, and the duties and responsibilities of every citizen; prepare maps of local facilities vital for civil defense, including telephone offices, police and fire stations, reception areas, and evacuation routes; and identify and study the structure of local, state, and national civil defense organizations. Biology students were to study the effects of nuclear radiation on animals; demonstrate how radioactive materials move from soil to plants; and report on the use of isotopes in combating disease. Other science activities included demonstrating the detection and measurement of radiation using a survey meter, dosimeter, and Geiger counter; constructing a model to illustrate a chain reaction; preparing reports on methods used in decontamination; and outlining the various ways of providing protection from fallout. In physical education, teachers should have their students demonstrate the types of exercises and games that can be used in shelters, while art students provided interior decorations for fallout shelters and designed civil defense-related posters. With the vital nature of women in a post-nuclear attack world, the handbook provided extensive activities for home economics students:

- Determine the kinds and amounts of food necessary for dietary requirements for an individual.
- Determine the necessary food for a given family in a shelter for a period of two weeks.
- Prepare basic menus for a two-week period of living in a shelter.
- Make a survey of additional items necessary to make the period in the shelter as comfortable as possible.
- Demonstrate a household medical and first aid kit.
- Outline the problems involved with the care of young children under emergency conditions.
- Discuss the role of "morale building activities" following a disaster. Describe several activities.
- Prepare a plan for caring for an evacuated family in your home.[8]

In Athens, Ohio, fifth-graders spent five weeks on a unit about atomic energy that also included discussions on the possibility of nuclear attack and protection from the effects of the blast. Rather than textbooks, the class used thirty-five books and booklets provided by the school librarian. Despite the emphasis on the effects of a nuclear attack, the children displayed "neither fear nor defeatism but faced this world problem in an alert and practical manner," according to an article in *The Elementary School Journal.* The class even wrote and mimeographed a nine-page illustrated booklet on atomic energy, and gave presentations to other classes on such topics as radiation, chain reactions, civil defense, and the disposal of radioactive waste.[9]

As President John F. Kennedy's New Frontier unfolded in the early

1960s—with his inaugural words still resounding: "Ask not what your country can do for you; ask what you can do for your country"—secondary and elementary school teachers continued their mission. This mission, which had spanned nearly two generations, had not wavered from instilling the knowledge and behaviors deemed essential for what many considered to be the eventuality of a nuclear war with the Soviet Union—an eventuality magnified by two confrontations in 1961 and 1962. The Berlin crisis, which occurred between June and November 1961, resulted from the Soviet Union's unsuccessful demand that Western armed forces be withdrawn from West Berlin and eventually led to the construction of the infamous Berlin Wall dividing the city into free and Communist-controlled sectors. A year later, in October 1962, the United States hovered on the edge of the nuclear war precipice over the installation of Soviet missiles in Cuba, just ninety miles from the nation's border, before Soviet Premier Nikita Khrushchev backed down.

Shortly after the 1961 Berlin crisis, New York University conducted a study to ascertain the opinions of children about the possibility of nuclear war. The children were asked just three questions: Do I think there is going to be a war? Do I care? What do I think of fallout shelters? Three thousand public and private secondary school students in three different regions—New York City, suburban Philadelphia, and upstate New York—participated.[10]

In response to the first question concerning the possibility of war, forty-five percent of junior high school respondents indicated they expected war, while senior high school students were slightly more optimistic. Concerning whether they care, nearly all children said, "Yes." Reasons varied, but the often fatalistic responses included the following: "I will die." "My parents, brothers, sisters, friends will die." "If my family dies and I live, I'd rather be dead." "Even if I survive, what will there be worth living for, with millions dead?" The final question about fallout shelters resulted in the widest discrepancy in responses. Among junior high students, forty-eight percent favored shelters; in contrast, only thirty percent of senior high students favored them. Those opposed offered these comments: "Radiation will seep in." "It's like building a tomb." "They are stupid, a farce, a money-making proposition." "They are just preparing people for a horrible war."[11]

During the first week of the Cuban missile crisis, the university conducted a follow-up study to a new group of 300 secondary school students with a similar background to the earlier study. Surprisingly, students responded more positively, with sixty-nine percent saying they did not expect war, compared to forty-eight percent not expecting war the previous year, and sixty-six percent of respondents voiced opposition to fallout shelters, a twenty-three percent increase from the 1961 study. The more children knew about the deadly potential of nuclear war, the study found, the less confidence they had in shelters. The decreasing confidence in fallout shelters, however, most likely reflected, at least in part, the wider public's declining faith in their protective value following the Soviet Union's test of a fifty-seven-megaton hydrogen bomb in October 1961.[12]

"The children of today's generation were not yet born on the day that a small, primitive A-bomb fell on Hiroshima and changed the world forever," wrote Milton Schwebel, a New York University education professor who oversaw the study. "They are, in fact, a unique generation—the first children in history to have lived their entire lives under the shadow of threatened thermonuclear extinction."[13]

Less than twenty years earlier, Americans, triumphant in the world's longest and most costly war, had entered what historian Tom Engelhardt labeled a "victory culture," which began in the fall of 1945 and continued well into the 1960s. These years witnessed economic growth, suburban expansion, exponential consumerism, the solidification of mass culture, and an unprecedented baby boom—with some four million babies born each year beginning in 1946 and totaling 77 million so-called "Baby Boomers" by 1964. Yet America's children coming of age during these years vacillated between enjoying the benefits of victory culture and confronting the horrors of the contrasting "nuclear culture." Children "held both the triumph and the mocking horror close without necessarily experiencing them as contraries," Engelhardt has written. "In this way, they caught the essence of the adult culture of that time, which—despite America's dominant economic and military position in the world—was one not of triumph, but of triumphalist despair."[14]

In 1962, Brock Chisholm, former director general of the World Health Organization, spoke to a special gathering of the U.S. Children's Bureau on the occasion of its fiftieth anniversary. In his address, titled "Children, AD 2012," he told the audience that if they helped the younger generation assume the responsibilities inherent in ensuring peace in the atomic age, there was still "reasonable hope" for the future. "It could be a place of freedom and human development well beyond our capacity for knowing now," he said. "Or it could be a place of utter misery and desperation with a few survivors scratching to make a living and to stay alive. And which of those will be the case? Because it is not probable that there will be some middle ground."[15]

Over the first two decades of the postwar era, the federal government, state governments, local communities, civil defense agencies, educational organizations, parents, and, ultimately, teachers, all strove to help elementary and secondary students avoid triumphalist despair and retain reasonable hope in the future by controlling their fear and trepidation about nuclear holocaust through such efforts as atomics. Yet by the mid–1960s, their efforts became supplanted by the civil rights movement and the Vietnam War, which had turned the nation's attention away from the possibility of a nuclear war to a more immediate reality. While thousands of young men and women battled hostile forces in Southeast Asia, thousands more took to the streets to protest what was happening to their country: racial discrimination and violent confrontations; anti-war protests led by young radicals; and the "peace movement" prompted by the aptly called counterculture. These youth, who had been exposed to the myriad conflicting and dreaded issues of the atomic age, had become engulfed in new issues perceived as threatening not civilization but rather society and their place in it. One could argue, perhaps, that atomics had, indeed, been successful. The Sixties Generation, those born in the 1940s and early 1950s and growing up with the constant threat of thermonuclear extinction, may not have had a united voice, but they clearly understood the necessity to promote brotherhood, protect democracy, and prevent war ... or else.

Conclusion

"All children are likely to be affected in some way by war or terrorism involving our country.... It is crucial to provide opportunity for children to discuss their concerns and to help them separate real from imagined fears."—National Association of School Psychologists[1]

In terms of the impact on school-age children, many similarities exist between the post–World War II era and the post–9/11 era of the twenty-first century. August 6, 1945, was a life-changing and world-changing historical event, as was September 11, 2001. The atomic bombings of Hiroshima and Nagasaki, Japan, brought an end to a deadly and protracted war that began in Asia with the Japanese invasion of China in 1937, and in Europe with the German assault on Poland in 1939. But the bombings also thrust the country into an unchartered atomic age replete with new fears and anxieties concerning the possibility of another war fought with atomic weapons.

Americans living at that time could remember an era before the Enola Gay, a specially equipped B-29 Superfortress, dropped the first atomic bomb on Hiroshima at 8:15 on a Monday morning, killing more than 70,000 men, women, and children instantly, and thousands more in the months and years to come—an era when national borders helped to defend against foreign enemies and, most important, helped to ensure a future. Children born in the mid–1940s and after, however, grew up knowing only a nation engaged in a heated Cold War against an atomic-armed enemy and being constantly reminded in school, at home, in the community, and in the media to be "alert today, alive tomorrow" or there would be no future.

Most Americans experiencing that tragic Tuesday morning in 2001 can remember where they were and what they were doing when American Airlines Flight 11, a Boeing 767 aircraft, flew into the North Tower of the World Trade Center in New York City at 8:46 a.m. Eastern time. They also can remember the end of the Cold War and a time when the nation seemingly had a bright and peaceful future. As with August 6, 1945, however, September 11, 2001, transcended the destructive acts committed that day and opened a new era where our enemy is no longer a single nation capable of inflicting nuclear annihilation but rather a nebulous array of rogue nations, foreign and domestic militant groups, and individual fanatics bent on destroying America in what is now a "War on Terror." Children born after 9/11 only know a world where terrorism can occur anywhere at any time by anyone—a similar situation encountered by school-age children in the 1940s and 1950s when the government and educators repeatedly warned that an atomic attack could occur anywhere at any time without warning. Moreover, the Federal Civil Defense Administration began operations in 1951, created specifically to prepare Americans for the probability of such an attack by the Soviet Union, and the Department of Homeland Security had its beginnings in September 2001, created specifically to address a nation facing a new enemy—and enemies—in what has become an unremitting war against militant terrorism abroad and random terrorist attacks at home.

In another response to 9/11, the United States Department of State, in collaboration with a committee of social studies educators, published *Terrorism: A War Without Borders*, a comprehensive resource containing video and print materials to help middle school and high school students better understand how world events are connected to their own lives and their community.[2] The resource, published in 2002, encouraged teachers to cover the following points:

- Terrorism has existed for centuries.
- Terrorists often challenge something, such as the actions or policies of a government, religious group, corporation, or other organization.
- There is no single definition of terrorism or who is a terrorist.
- Some terrorist organizations:
  o Are recent while others have a long history;
  o Have many members while others have just a few members;
  o Are well-organized while others have an undefined structure;
  o Are well-financed while others have meager resources.

- Terrorists and terrorist organizations often:
  o Commit public acts of violence;
  o Create fear and apprehension among the populace of an area;
  o Seek to attract publicity for their cause;
  o Attack civilian rather than military targets;
  o Claim political motives for their actions.
- Responses to terrorist activities can include:
  o Statements condemning their actions;
  o Initiatives to obtain international cooperation to prevent terrorist acts and punish terrorists;
  o Political or economic sanctions against the region of their home base and/or the nations supporting them;
  o Criminal prosecution;
  o Military action against the terrorists and their supporters.

Teachers had the flexibility of using the suggested lessons, which reflected *Expectations of Excellence: Curriculum Standards for Social Studies* developed by the National Council for the Social Studies, as a separate unit, or they could incorporate the array of materials provided in the publication into their existing lesson plans—as teachers had incorporated government materials related to the atomic bomb into their plans in the late 1940s.

In addition to the federal government's efforts to provide post–9/11 educational materials, states also revised their educational standards to meet the need to prepare students for the War on Terror. Louisiana, for example, now requires the teaching of both domestic and foreign terrorism; students in Massachusetts learn about the rise of Islamic fundamentalism in the last half of the twentieth century; Michigan students analyze the causes and challenges of continuing and new conflicts by discussing tensions resulting in ethnic, territorial, religious, and nationalistic differences; in Virginia, students must demonstrate knowledge of the nation's foreign policy since the end of World War II; and the State of Washington requires students to weigh the argument that 9/11 is the sole cause of the War on Terror, and to understand the ramifications of the Patriot Act, which gives the government the right to search or surveil homes and businesses for suspected terrorist activities.[3]

Unfortunately, fifteen states mention terrorism or the war on terror but do not identify the 9/11 attacks, and another fourteen states fail to include any mention of 9/11 or terrorism. A 2011 study, published by the *Center for Information and Research on Civic Learning and Engagement* at

Tisch College, Tufts University, also found that although 9/11 is a commonly required topic, the actual events of that day and their context are described only briefly in recent textbooks.[4] Conducted by Professor Jeremy Stoddard, College of William & Mary, and Professor Diana Hess, University of Wisconsin–Madison/Spencer Foundation, the study focused on the representation of 9/11 and terrorism in curricula, textbooks, and state standards documents. The authors found textbooks published soon after 2001 form a consensus that the 9/11 attacks were, indeed, unprecedented and of significant historic importance. Yet textbooks, as well as school curricula, published later in the decade lack opportunities for higher-level discussions. Rather, the emphasis shifted to basic reading comprehension while failing to stimulate students to analyze, synthesize, or construct new concepts. Very simply, more recent textbooks stressed the memorization of facts and a cursory understanding of the events of 9/11.

By comparison, supplemental curricula provided by nonprofit organizations include more in-depth information and multiple perspectives on terrorism-related issues such as how democracies should strike the right balance between security and civil liberties. The 4 Action Initiative, one such supplemental program, released a curriculum in 2011 to assist New Jersey teachers in educating students about the threat of terrorism.[5] A collaboration of Families of September 11, Liberty Science Center, and The New Jersey Commission on Holocaust Education, the 4 Action Initiative's *Learning from the Challenges of Our Times: Global Security, Terrorism and 9/11 in the Classroom* offers lesson plans for students in elementary school, middle school, and high school. The plans cover the same topic areas, with the depth of information increasing at each successive grade level. Topics include human behavior; violence, aggression, and terrorism; the historical context of terrorism; a contemporary case study in terrorism; challenges and consequences in a post–9/11 world; remembrance and the creation of memory; and building better futures. For example, students in kindergarten through fifth grade learn about hurtful words and how to respond to them. In grades seven and eight, the discussion advances to the meaning of terrorism and examining different cases of terrorism. High school students focus on the definition of a terrorist and terrorism, including discussions of terrorist acts, international law, and the judicial system. Teachers are encouraged to create a safe and comfortable classroom environment, to present information in a calm and reassuring manner, to avoid traumatic images of destruction, to encourage positive school and community projects, and, most important, to minimize their students' fears—the same instructions teachers received in the Cold War era.

The Federal Emergency Management Agency (FEMA), in cooperation with the American Red Cross and the U.S. Department of Education, also has released a strategy to educate children about disaster prevention, protection, mitigation, response, and recovery.[6] The *National Strategy for Youth Preparedness Education: Empowering, Educating, and Building Resilience* (National Strategy), published in 2014, focuses on preparing young people to be ready for any disaster, natural or manmade. Echoing the same concern from more than a half century earlier, the National Strategy concludes that although children are disproportionally affected by disasters, they are not prepared because of insufficient attention. Even though FEMA does not use the term terrorism, it is understood that youth must be prepared for any disaster, which reflects the same comments by government agencies beginning in the late 1940s. Additionally, FEMA argues that children who learn about emergency preparedness experience less anxiety during an actual disaster.

Reflecting the atomics approach, the Public Broadcasting System (PBS) developed a teacher's guide titled *Roots of Terrorism*, recommended for use by teachers of high school civics, geography, global studies, world history, language arts, economics, and United States history.[7] Curriculum standards for civics students stress the importance of understanding the role of diversity in American life; disparities between ideals and reality in American political and social life; the formation and implementation of public policy; the organization of nation-states and issues concerning U.S. foreign policy; and the impact of political and nonpolitical developments in the U.S., as well as other nations. History students must understand developments in foreign policy. Economics students learn the basic concepts of international economics. And geography students discuss why and how people's perceptions of places and religions differ.

The guide is based on four films, each linked to a website with extended transcripts with government officials, other policy experts, and journalists, as well as additional information such as chronologies of key events and extended analyses of the film's topic. The films are *Hunting bin Laden*, an investigative report on the man who declared holy war on the United States; *Target America*, tracing the terrorist war against the nation from the early 1980s to the present; *Looking for Answers*, examining the roots of Islamic terrorism; and *Saudi Time Bomb?*, an examination of the relationship between the U.S. and Saudi Arabia.

Various activities are recommended, including one on defining terrorism, which encourages students to recognize that people disagree on who should or should not be labeled a terrorist; to define the words "terrorist" and "terrorism" and apply the definitions to past and present events; to identify discrepancies in the nation's foreign policy as it relates to terrorism; and to examine the connection between terrorism and human rights.

PBS continues to provide teachers with resources that can be incorporated in lesson plans for each grade level:

- Pre-K through Grade 5—"The American Flag" provides the history of the flag, what it represents, and the proper way to display it.
- Kindergarten through Grade 5—"A Nation of Many Cultures" encourages students to create a visual history of their family heritage, then discuss similarities and differences with other students.
- Middle School—"Emergency Preparedness" introduces students to governmental and humanitarian responses to natural and man-made disasters.
- High School—"Taming Terrorism" examines how international organizations are working to eliminate terrorism.

As PBS and the other educational organizations underscore, terrorist attacks in our country and threats or realities of war are frightening experiences for all Americans, but especially to school-age children. Yet the message being articulated is not new.

"Terrorism is insidious in infiltrating the collective psyche with fear and the pervasiveness of our horror," wrote educators Ilene R. Berson and Michael J. Berson just one month after 9/11. "Children and young adults are especially vulnerable to the psychological impact.... Acts of terrorism instill fear and helplessness in a society. Coping with the intense stress and trauma of these events can be overwhelming for our youth, who may feel especially vulnerable. To function optimally, each student has a basic need for safety and security."[8] A month following the atomic bombings of August 1945, *Senior Scholastic* magazine wrote, "The atomic bombs that fell on Hiroshima and Nagasaki did more than devastate those cities and shorten the Pacific war.... The atomic bomb is bigger than the Big Five. Its power is greater than the combined armed forces of the United States, Russia, Great Britain, China, and France. If we permit another war, it will destroy us all."[9]

From the mid–1940s through the 1960s, in an effort to help school-age children cope in a precarious atomic age, schools, with the assistance of federal and state government agencies and educational organizations, adopted the atomics curriculum in all classrooms at all grade levels. Today, educators, with the assistance of government, nonprofit, and educational agencies and organizations, remain committed to preparing school-age children to cope in a terrorist-infected world. Whether it is the threat of atomic annihilation and or the threat of random terrorism, the mission for teachers does not waver; rather, it is always to help students understand the issues, the implications, and the impact of the threat in order to be prepared for a situation that may occur without warning.

Appendix A

Suggested Learning Experiences

In December 1946, the University of Illinois Bulletin *devoted its entire issue to "Living in the Atomic Age: A Resource Unit for Teachers in Secondary Schools," compiled by a committee of nine educators, including a superintendent of schools, a high school principal, four high school teachers, a librarian, and a junior college teacher, under the direction of Harold C. Hand, a professor of education at the University of Illinois. The resource unit resulted from a request by the National Committee on Atomic Information for the university's College of Education to design a program to help teachers "discharge what is probably the most urgently critical task with which the schools of America and the world have ever been confronted:" life in the atomic age. The following excerpt offers suggested learning experiences for secondary school pupils. Reprinted with permission from the Bureau of Educational Research at the University of Illinois in Urbana-Champaign.*

NOTE TO THE TEACHER: If you have had little or no experience in using a resource unit ... the important thing to be remembered is that this is *not* a teaching unit. Rather, it is a storehouse of sorts to which you can fruitfully turn for vital and workable suggestions in building a teaching unit tailored by you to fit your particular class group. In building such a unit, you would, of course, select from among the following activities only those which you believe to be appropriate for this purpose.

A. Initiatory Activities

1. Have book jackets from materials dealing with nuclear energy, the atom bomb, need for world government, and related matters displayed on bulletin boards and in classrooms.
2. Utilize a modification of the Lucky Strike radio curiosity arouser "LS/MFT" device on bulletin boards and in classrooms. Examples: "IA/MWW3" (International Anarchy Means World War III). "WT/MAB" (War Today Means Atom Bombs).

"AW/MDC" (Atomic War Means the Destruction of Civilization). "AB/MWG" (Atomic Bombs Mean World Government).

3. Display selected pages from the "atomic bomb" issue of *Look Magazine* on bulletin boards. Reprints are available for a few cents from the National Committee on Atomic Information, 1749 L Street N.W., Washington 6, D.C.; a copy is also included in the "Study Kit on Atomic Energy," which is available for $1 from this organization.
4. Examine recent issues of other magazines for pertinent pictorial materials to display on bulletin boards.
5. Show the cartoon film strip, "How to Live with the Atom." This and the accompanying dialogue may be secured for $2.50 from the National Committee on Atomic Information (see above).
6. Display newspaper cartoons dealing with the atom bomb and related questions.
7. Ask your local Kiwanis, Lions, Rotary, or other service club to secure a good outside speaker on some question related to the Atomic Age and to loan him to the school for a talk to the pupils. The NCAI (National Committee on Atomic Information) maintains a roster of such speakers, some of whom may live near your locality. Speakers on such subjects can also usually be obtained from the nearest college or university.

B. "Finding Out About It" Activities

8. Stimulate the pupils to collect, report, and discuss the statistics regarding the destruction at Hiroshima and Nagasaki *(One* World *or None, Must Destruction Be Our Destiny? Dawn Over Zero,* and John Hersey's article in *The New Yorker* are especially good sources for this purpose. Through this discussion, help the pupils understand why the atom bomb is called the "weapon of saturation." Also invite them to reflect upon the fact that but one atom bomb (each quite primitive and now obsolete) exploded in each of these destroyed cities, and that it is improbable that single atom bombs will ever again be used should there be another war.
9. Encourage the pupils to apply their findings from their study of Hiroshima and Nagasaki to their own community or to a neighboring city. One way to do this is to have them prepare two maps of the community or city in question and mark on one the spots of an imaginary atom bomb explosion or explosions. Hospitals, fire stations,

newspaper offices, radio stations, power stations, water works, railroad stations, factories, business areas, residential districts, schools, parks, thoroughfares, etc., should be plotted on one of these maps—the "before" map. On the other, the area or areas of probable destruction should be charted by means of pins, sketches, and the like. This last map should be translated into words, figures, graphs, etc., giving the details of this probable destruction (the references noted in "8" above will also be helpful for this purpose). The probable effects upon those aspects of social and economic life most familiar and vital to the pupils concerned should be emphasized in the discussion.

10. The second map noted in "9" above might well be superimposed by the pupils upon a wall map of the surrounding territory. This larger map should show the types of agriculture, forestry, manufacture, etc., which are related to markets, banks, factories, stores, and so on, in the city. The probable consequences of the imaginary destruction in the city should then be charted, made explicit in figures, graphs, and words, and discussed by the group.
11. The pupils might also be encouraged to carry out a number of projects similar to "9" above, but in reference to the particularly crucial cities of the U.S.A. (New York, Chicago, Detroit, San Francisco, Pittsburgh, St. Louis, Washington, D.C., for example). In this regard, the pupils should be led to list and discuss the probable effects upon their lives and those of the rest of the country's population if our great centers of finance, manufacture, shipping, and government were destroyed.
12. The consensus of the pupils' opinions which derive from the projects noted above might well be expressed by some members of the group in a dramatic skit entitled "Life in _____ One Year After an Atomic War." Other pupils may wish to present cartoons, sketches, themes, poems, or articles for the school and/or community newspaper centered around the same theme. Still others may prefer to express their ideas in this respect in "before" and "after" models in wood or clay.
13. Lead the pupils to study and discuss the results of the Bikini atom bomb tests. In particular, have them contrast what the official boards (the President's Civilian Commission and the Army-Navy Joint Chiefs of Staff) actually reported and the generally erroneous opinion which most people appear to hold regarding the destructiveness of these test explosions. Again, be certain that the pupils take into account in their discussion the probability that several, instead of

single, atom bombs will be used in any attack should we permit another war to occur.

14. If one or more of the above projects has been carried out, there is certain to be a great deal of pupil curiosity regarding the power resident in the atom. Have the pupils collect photographs and statistics and make graphs or charts showing the difference in explosive power and in cost (see General H. H. Arnold's chapter in *One World or None*) between the conventional and the atom bomb. Encourage the pupils to make charts or graphs to compare nuclear energy with other forms of energy. Have them actually compute energies and convert from the c.g.s. system to the English system. For example, apply the formula $E = mc^2$ and show how many tons of coal would have to be burned to produce the equivalent of the nuclear energy in an ounce of this fuel. (If you are not a teacher of mathematics, you might ask a colleague in this field to help you out.) Have the pupils consider what the relative cheapness of atomic explosives means in terms of the changed ratio between the potential warring powers of "big" and "little" nations.

15. Direct the pupils in compiling a glossary of "atomic terms" in pupil-formulated definitions. All the words commonly used in newspaper, radio, and other discussions of atom bombs, nuclear energy and related matters should be included in this list. The formulating of this list might well be a whole class project. If you are not a science major, you should probably have your pupils consult some teacher in this field for verification of their definitions.

16. Many of your pupils will probably want to construct models of atoms; all will be helped by having such models before them. For simpler atoms, sponge rubber balls or wooden balls can be used to represent the nuclei and electrons. Strands of wire can be utilized for mounting and for representation of electron orbits. For more complex atoms, have the pupils make representative drawings. Example: U-235 and possible fission products.

17. Encourage the pupils to prepare disintegration charts for radioactive elements, showing which particles are released and what elements are left from the reaction. Have them trace the steps all the way from uranium to lead. Also have the pupils show the integrative steps necessary to make neptunium and plutonium.

18. Explain to your pupils how the fission process is apparently induced in the atom bomb; discuss this sufficiently to insure that the elementary principle involved is understood by all members of the group.

19. Have the pupils read about and discuss the various methods by which the U.S. Government produced fissionable materials in the atom bomb project. Charts, diagrams, and photographs will prove helpful in this undertaking.
20. Suggest to the pupils that they examine the numerals on a luminescent watch dial under a magnifying glass in a dark room. Have them observe the individual flashes of light and tell what causes them. What radioactive substance is present?
21. Encourage the pupils to make a chronological table of the scientific developments leading to the discovery of fission. Have them make a similar table showing the atomic developments in this country between the time of the fission discovery and the time atom bombs were produced. See to it that the names and nationalities of the people who made the discoveries are included by the pupils in both of these tables. Then have them discuss the question of the degree to which these discoveries were "purely American."
22. One outgrowth of "21" above might well be the making of a "Who's Who" booklet.
23. Another desirable outgrowth might well be the construction of a time line chart showing the development of nuclear energy. Have the pupils choose a small enough scale so that the scientific acceleration due to the war may be clearly shown. This graph might also be projected by the pupils to afford some estimate of the probable rate of future developments.
24. Still another desirable outgrowth of "21" might be to have the pupils make a list of the important persons whose talents were lost to Germany by her policy of discrimination and suppression. Encourage the pupils to speculate as to what bearing this probably had on the outcome of the war, but insist that they give their reasons for their estimates.
25. Suggest to the pupils that they collect and discuss the opinions of atomic scientists concerning the question of whether or not America possesses some essential basic scientific secret which scientists in other countries do not know, and without which their governments cannot manufacture atom bombs. Also have them collect the opinions of these scientists in reference to the time it will take other nations to begin producing atom bombs.
26. Ask the pupils to look up, report, and discuss what atomic scientists have to say about the probabilities of sabotage within our own borders should the atomic armament race continue, noting particularly

the counter-measures which are believed to be necessary and how these would destroy various personal freedoms which Americans treasure.

27. Encourage the pupils to poll the published opinions of the atomic scientists and of our military leaders in reference to whether or not any adequate scientific or military defense against the atom bomb is likely within the foreseeable future. Have them list and discuss the implications of what they will thus discover—namely, that no scientific or military defense is anticipated by these experts, that the only defense they can envisage is political (i.e., world government, at least enough to control fissionable materials), and that the only other alternative is a policy of decentralization (of cities, industry, government agencies, universities, etc.). Help the pupils to envisage what this last alternative would entail—time required, cost, loss of personal freedoms, effects on real estate values, effects on commerce and industry, etc.
28. Ask the pupils to prepare reports outlining what various organizations and individuals are doing to prevent an atomic war. Have them consult the bulletins of the National Committee on Atomic Information, the *NEA Journal,* and whatever guides to periodical literature the school and community libraries may contain. Do not overlook the work of the Campaign for World Government (505, 343 South Dearborn, Chicago, Illinois), of Americans United for World Government, Inc. (1860 Broadway, New York, N.Y.), or of Chancellor Hutchins's group at the University of Chicago. Also have the pupils read and discuss the press and other criticisms of the work of these organizations and individuals.
29. Suggest that the pupils make a survey of newspaper and other editorials concerning the atom bomb, nuclear energy, world control of fissionable materials and related matters. Encourage them to evaluate these editorials in terms of what the effects on world peace would probably be were the action either expressly urged or implied in each translated into government or other policy.
30. Ask the pupils to collect, display, and evaluate the probable effects of newspaper and other cartoons treating some aspect of the atom bomb, nuclear energy, world government, and related matters.
31. Encourage the pupils to consult the radio guide section of some city newspaper for the past four or five months, note the programs which dealt with some phase or aspect of the atomic crisis, and write to the broadcasting companies concerned for mimeographed

copies of these broadcasts. Parts could then be assigned to each of the members of a committee or of the class and the broadcasts thus read to the total group. This should, of course, be followed by discussions in which the pupils would evaluate what was either explicitly or implicitly urged by the broadcast.

32. Suggest that the pupils make either a group or individual scrapbook of current information (articles, editorials, pamphlets, photographs, cartoons, etc.) concerning the atom bomb, nuclear energy, world government, and related matters. The scrapbook might contain such section headings as the following: Technical Developments; Results of the Hiroshima and Nagasaki Bombings; Results of the Los Alamos and Bikini Tests; The Atomic Scientists and Their Contributions; Nuclear Compared with Other Forms of Energy; Proposals for the Control of the Atom Bomb; Peace-Time Uses of Nuclear Energy. The scrapbook should be loose-leaf in order that it may be more easily kept up-to-date.

33. Suggest to the pupils that they prepare, keep up-to-date, and make available in some popular spot in the library a "Guide to Reading Materials" in which they list with brief annotations current articles, editorials, pamphlets, and books bearing on any aspect of the Atomic Age. For the most part, these listings should be from materials which are available in the local school and community.

34. Ask the pupils to prepare, keep up-to-date, and post conspicuously a "Guide to Coming Broadcasts" which deal with any aspect of the atomic question. The Sunday editions of metropolitan newspapers and the bulletins of the National Committee on Atomic Information will prove very helpful in this regard.

35. Have the pupils prepare, keep up-to-date, and post conspicuously current cartoons, photographs, and other pictorial materials which treat of any aspect of the atomic problem.

36. A committee of the class might well be stimulated to call regularly upon the managers of the local cinemas for purposes of preparing a continuing weekly or monthly list of "Coming Movies" (newsreels, shorts, other features) dealing with any phase of the problem of nuclear energy. The titles on this list should, if possible, be briefly annotated and should, of course, indicate time, place, and price. Obviously, it should be posted in some conspicuous spot.

37. Direct the pupils in preparing a skit in the form of a radio quiz program or "Man on the Street" interview in which members of the class present the views of prominent men by impersonation.

This could be presented in assembly and before lay adult groups as well as before the class.
38. Some members of the class might be stimulated to write a play concerning racial discrimination and intolerance to show the adverse relationship of these special forms of ignorance to world peace in the Atomic Age. This might then be given before assembly and lay adult groups as ·well as before the class.
39. Stimulate the pupils to conduct symposiums or panel discussions on the following questions related to world government:
   a. By definition, what does the term "unlimited national sovereignty" mean?
   b. To what degree do national states today have effective control over what goes on within their respective boundaries in reference to such things as:
   (1) Prosperity.
   (2) Depression.
   (3) Currency regulation.
   (4) Armaments.
   (5) Being involved if war breaks out.
   (6) Epidemics.
   c. To what degree is "effective unlimited national sovereignty" in reality a myth in today's world?
   d. To what degree does the attempted practice of unlimited national sovereignty spell international anarchy today? Regardless of the consequences to other nations, what things is each national state legally free to do today?
   e. Is continuing peace possible or likely in a world of closely interdependent national states, given the present facts of international anarchy?
   f. Reves (see his *Anatomy of Peace*) enunciates the principle that war occurs when non-integrated social units of equal sovereignty clash in their interests; that peace is assured only when these units are integrated under a higher order of sovereignty. Does this appear to have been historically true in reference to the resolving of wars among clans and among city-states? If so, what is implied in reference to the necessary machinery for peaceably resolving the present clashes of interests among today's 70 or more non-integrated national states of equal sovereignty?
   g. How long would peace obtain within the U.S.A. were there no laws definitive of rights and responsibilities and backed up by a

police power sufficient to coerce offenders? Is it reasonable to expect world peace and the control of atomic energy under similar conditions?

    h. Actually, under effective world government, would any given national state (our own, for example) gain or lose in its real power to control the actual conditions which result in prosperity or depression, peace or war, for its citizens?

40. Ask the pupils to consider what part America's great industrial strength played in the defeat of the aggressor nations in World Wars I and II. Then have them discuss the question of whether or not these two lessons of history are likely to be overlooked in the event of another war. What nation would in all probability be the first to be attacked should World War III be permitted to occur?

41. Lead the pupils to make a study of the revolutionary effects of former inventions (for example, the automobile, airplane, radio, steam engine, etc.) whose consequences were *not* anticipated and planned for by social scientists, statesmen, and educators. Which of the evil effects could probably have been prevented or softened, and which of the benefits more widely enjoyed, had such prophecies been attempted and had they resulted in the indicated social action (laws, changed school curriculum, etc.)? Ask the pupils to discuss what they believe the moral to be so far as the new discovery of nuclear energy is concerned.

42. Let the pupils interview four or five people in each of the following occupations and ask them what they think the effects of an uncontrolled atomic armaments race is likely to have on people in their respective lines of work: merchandising, real estate, the ministry, mechanics, newspaper work, salesmanship, scientists, utilities work, and teachers. Then have the pupils compare these answers with the predictions in this regard given in the reprint from *Look Magazine* which is contained in the Study Kit on Atomic Energy of the National Committee on Atomic Information. Have the pupils indicate what they conclude from these two findings in reference to the need for adult education in their community.

43. Let the pupils conduct a carefully planned poll among the adults and high school youth of the community to discover their attitudes toward keeping the "secret" of the atom bomb, how long they think it will take other countries to produce such bombs, whether or not there is likely to be an adequate scientific or military defense against the bomb, and whether or not fissionable materials should be con-

trolled by some type of world government. See to it that the pupils review their findings against the background of what the atomic scientists, military leaders, educators, and other men prominent in American life say about these matters. Have the pupils note what, if any, needs for community-wide education this review brings to light.

44. Interest the pupils in studying and discussing the historical development of attempts at international cooperation to date. This should include the Hague Conference, the League of Nations, the Kellogg Plan, the International Food Conference, Bretton Woods and its outcome, U.N.R.R.A., U.N., and U.N.E.S.C.O. What did, or does, each attempt to do? What was, or is, being accomplished? What were, or are, their successes and failures—and why?

45. Suggest that the pupils compare and contrast the League of Nations and the United Nations. Who was, or is, represented? How was, or is, each structured? To do what? By what was, or is, each most seriously limited? Was either intended to yield real world government under true international law?

46. Have the pupils list and discuss the present barriers to international understanding, reciprocal good will, and world government under true international law. What and where are the most crucially needed natural resources, and who controls them? In what other respects, and where, do the economic interests of various national states clash? How do the unflattering and inaccurate stereotypes which each national group has of the others affect the situation? The self-assumed superiority of each national group over the others? The ignorance on the part of each national group of the history, traditions, mores, folkways, customs, hopes, fears, frustrations, problems, and needs of other national groups? The lack of an international history? Language barriers? The assumption of "white superiority"? Differences in economic and political systems? The assumption of self-righteousness on the part of all national groups, with the consequent unwillingness to compromise? Insecurity? Fear of political, economic, or military aggression?

47. Direct the pupils to read pp. 1–23 in Reves, *The Anatomy of Peace*, to note how very differently the peoples of various national states conclude from the same set of objective facts. Then have them try their hand at describing how various recent events (such as the Bikini tests, Russia's behavior at the Peace Conference, the behavior of the U.S.A. at the Peace Conference, the U.S.A.'s claim to island

bases, Britain's behavior in Palestine, and so on) probably appear to the respective peoples of the major national states of the world.

48. In the light of the above, have the pupils estimate how the people of the U.S.A. probably appear to other national groups, particularly to the Russians. The question should be raised as to whether we would behave any more rationally than do the Russians were our situations reversed, i.e., were they in effective possession of island bases a few hundred miles from Los Angeles, San Francisco, and Seattle; were they in exclusive possession of the atom bomb; had they destroyed two cities with such bombs; had they recently still further demonstrated the awful destructiveness of this weapon; were they presumably continuing to stock-pile atom bombs; were we a great industrial power with no warm water port; had they the greatest industrial machine on earth; had they the greatest navy and air force; were they said to be training and equipping an army in a country immediately adjacent to ours; had they sent troops into our country to coerce our people as recently as 1919–1921. The raising of such questions should in no way imply a "whitewash" of any and all Russian actions; instead, the pupils should be encouraged to be properly critical of any threat to world peace, Russian or otherwise. But if war now means atomic war, if atomic war means destruction, and if peace can be maintained only on the basis of mutual understanding and reciprocal good will, it is essential that we estimate how we look to the other fellow, and vice versa. And if what we see when we look at ourselves from the other fellow's point of view is not altogether flattering, then the thing to do is to make amends or change our ways—not bury our heads in the sand.

49. Help the pupils to make a study of our stereotypes ("pictures in the head") of other nationals. For example, the "British stereotype." Have them collect cartoons and report on radio programs; motion pictures, and vaudeville programs in which Englishmen are portrayed. Then have them list the adjectives which best describe this portrayal, and note how few are complimentary in character. Then, if possible, invite somebody who really knows England and Englishmen to comment on this list—most of the items will probably be false as far as most Englishmen are concerned. If an Englishman can be secured, ask him to characterize the way in which Americans are portrayed in British vaudeville, cartoon strips, and popular opinion. This stereotype will likewise be mostly false so far as most Americans are concerned.

50. Let the pupils investigate the probable peace-time uses of nuclear energy which promise to benefit mankind. This investigation might well be organized under the following headings:
   a. Medical uses.
   b. Heat—to counteract cold weather; to generate electricity, etc.
   c. Power for transportation.
   d. Power for manufacture.
   e. Saving of coal and petroleum for use in synthetics.
   f. Speculations regarding changing the weather.
   g. Speculations regarding the direct synthesizing of the elements (i.e., the by-passing of pastures and sheep in producing clothing, of corn and pigs in producing food, of forests and saw mills in producing shelter, etc.).
51. Individual pupils with active imaginations might be asked to prepare themes, sketches, cartoons, or dramatic sketches showing how the application of nuclear energy to peace-time pursuits may change our occupations, our fight against disease, our food, clothing, and shelter, our recreations, our transportation, our home life, our government, etc.
52. Lead the pupils in discussing the benefits likely to accrue if scientists the world over are free to carry on nuclear research related to beneficial uses of this new type of energy. Have them note how nuclear developments to date are international in scope, and estimate how seriously progress will be impeded in the future if scientists are forced to work in isolation from one another, and in secrecy, under conditions of military security. Also see to it that the inherently international character of science is understood and appreciated (Wylie's "Blunder" will be found helpful here).
53. Suggest that the pupils study and discuss the ways in which science, invention, and modern technology (in industry, transportation, communication) have made the peoples of the earth closely interdependent and effected a virtual revolution of which most people are too little aware. (Staley's *World Economy in Transition* and Chapter I in Melby's *Mobilizing Educational Resources* are especially helpful in this regard.)
54. Have the pupils conduct a round-table discussion of the question "Are there today any important national problems which are not also international in scope—i.e., the resolution of which do not affect the well-being of peoples in other national states?"
55. Interest the pupils in studying and discussing what social scientists,

especially anthropologists (e.g., Ruth Benedict, Margaret Mead) have to say about the innate capacities of racial and national groups. Then have them consider the costly consequences in today's world of the special form of ignorance which leads each national group, our own included, to assume that it is innately superior to the rest.

56. Suggest that the pupils study the various great religions of the world with a view to discovering what each teaches in reference to "being your neighbor's keeper," "the brotherhood of man," etc. The purpose here should be to appraise the degree to which each is in harmony with the idea of world government.

57. Watch the press for an announcement of the proposed structure for world government on which Chancellor Hutchins and a committee of distinguished professors at the University of Chicago and elsewhere have been at work since November, 1945. When this appears, secure one or more copies and have these studied and discussed by the pupils.

58. Have the pupils study and discuss the plan for the world control of atomic energy proposed by the Committee on Atomic Energy of the U.S. Department of State (see Bibliography). Especially, have the pupils list and discuss the basic principles and the key provisions here put forth.

59. Be alert for all news items and other information concerning the recently (July, 1946) authorized Atomic Energy Commission. See to it that the pupils keep themselves abreast of its work through reading, radio listening, and class discussion.

60. Let the pupils study and discuss the charter provisions and the continuing work of the United Nations, of the United Nations Educational, Scientific, and Cultural Organization, and of the United States Atomic Energy Commission.

61. Secure materials descriptive of the purposes and program of the division of International Educational Relations of the U.S. Office of Education and have these studied and discussed by the pupils.

62. Suggest that the pupils write to the President of the Youth Committee on the Atomic Crisis of the Oak Ridge, Tennessee, High School for information concerning the significant educational work the Y.C.A.C. has done and is doing. Then have the class decide which of these activities suggest things which it could and would like to do. (The Y.C.A.C. represents a project which originated in the English classes of Mr. Philip E. Kennedy.)

C. "Doing Something About It" Activities

63. Direct the pupils to list the organizer groups in the community which support educational programs of any sort (service clubs, labor unions, P.T.A.'s, churches, etc.) and then have them suggest to the program chairmen of these organizations that they each schedule one or more meetings devoted to the problems of atomic energy. These suggestions could be conveyed by small committees of the class or by letters written by the pupils. The class might be able to "borrow" some of the speakers thus utilized and have them speak at school assemblies. Further, it might be suggested to these community organizations that one or more pupils from the class be invited to be present at the suggested meetings as reporters for the class.
64. Interest the pupils in preparing talks or dramatic skits on various of the topics or problems noted in Section "B" above and have these presented before as many interested community organizations as possible. These could also be included in assembly programs in your own and other schools, and given before other class groups.
65. Suggest that the pupils write articles or editorials for the school paper. This might take the form of a weekly column devoted to current atomic and related developments.
66. Let those pupils who have the knack draw cartoons dealing with atomic problems for the school paper or for posting on bulletin boards. All members of the class might well assist in formulating the ideas for these cartoons.
67. Suggest that the pupils prepare posters calling attention to forthcoming radio programs devoted to various phases of the atomic problem. This information might well be supplied to the school and community newspapers.
68. As the pupils uncover particularly significant articles or other printed items dealing with problems of the atomic age they might attempt to induce the editors of local newspapers to reprint them for the education of the public.
69. It should be possible for the pupils to put on a round-table discussion on local radio stations from time to time concerning problems of living in the Atomic Age which they have studied in class. Similar broadcasts might also be given over the school's public address system.
70. Suggest that the pupils arrange for an exhibit space in the window of the public library, a store, newspaper office, etc., for displays of materials (books, posters, photographs. cartoons, notices, etc.) bear-

ing on atomic problems; these displays the pupils should plan, prepare, set up, and maintain.
71. Letters to the Editor might be prepared by the pupils giving their views and questions concerning various aspects of the atomic problem.
72. Letters to Congressmen expressing the views of the group in reference to the control of nuclear energy and related matters might also be written by the class.
73. Arrangements might be made through the Junior Red Cross, or through the Division of International Educational Relations of the U.S. Office of Education. for correspondence with class groups in European, African, Asiatic, South American, New Zealand, Australian, Scandinavian, and other countries in reference to atomic control and related problems.

D. Evaluative Activities

74. At the beginning of the unit have pupils list the questions which they would like to have answered concerning the atom bomb, nuclear energy, and related problems. At the conclusion of the unit, or at any desired time in process, have the pupils list the questions the answers to which they feel every effective citizen in the Atomic Age must have or be attempting to secure. Then compare—better yet, have pupils compare—these two sets of questions to note evidences of (a) growth and (b) "blind spots" which still remain.
75. Have the pupils draw up a list of generalizations or conclusions which grew out of their study—items variously touching upon as many as possible of the types of learnings suggested in Part "B" above. These should be as simple and specific as possible and include only what the pupils are prepared to defend. At whatever time or times this evaluative device is utilized it will afford a reasonably valid indication of the pupils' status of development in reference to the area concerned.
76. File dated pupil work samples as the unit progresses and scrutinize these for evidences of growth.
77. File dated anecdotal accounts of significant things noted in pupils—evidence of a fact grasped, a generalization formulated and accepted, etc. Examine these for indications of growth.
78. As a summarizing activity, have the pupils compile a handbook on atomic energy. This should include in brief form what the pupils

believe to be the minimum essentials of factual information, the principal reasons why the problem is of real concern to everybody, the conclusions or generalizations which they have formulated, and the unanswered questions or problems of which they are aware. In addition to affording significant evaluative evidence, this document will be of real value to other pupils if put to use in other classes or in the school library.

79. Develop an inventory of attitudes, administer it at the beginning and again at the conclusion of the unit, and contrast the results. Such an inventory should contain items touching upon all of the types of considerations to be dealt with. The following might perhaps serve as samples:

    (1) How do you regard the atom bomb? (Check *one*)
        _____ a. It has no special significance (meaning) except that it ended the war with Japan.
        _____ b. It is important only as a new weapon of warfare which is of little concern to the average man.
        _____ c. It is a weapon which is so destructive that the majority of the civilians on both sides would be killed or incapacitated in an atomic war.
        _____ d. I have no opinion concerning the atom bomb.

    (2) Will war always be inevitable (unavoidable) because it is human nature to fight? (Check *one*)
        _____ a. Yes.
        _____ b. No.
        _____ c. I don't know.

    (3) Are Americans innately superior (born superior) to other people? (Check *one*)
        _____ a. Yes.
        _____ b. No.
        _____ c. I don't know.

    (4) Are white people innately superior (born superior) to colored people? (Check *one*)
        _____ a. Yes.
        _____ b. No.
        _____ c. I don't know.

    (5) How would you describe the *average* Englishman? (Check *all* statements which tell what you think.)
        _____ a. He has a poor sense of humor.
        _____ b. He is "stuck up" or "high hat" or "snooty."

_____ c. He is cold and unfriendly.
_____ d. He doesn't pronounce words the way they should be spoken.
_____ e. He likes to suppress other people.
_____ f. He isn't as intelligent as most Americans.
_____ g. He won't pay his debts.
_____ h. He wants everybody else to do his fighting for him.
_____ i. He doesn't make as good a soldier as the average American.
_____ j. He isn't democratic.
_____ k. The average Englishman is about as nice, as able, and as democratic as the average American.
_____ l. I don't have any opinion about the average Englishman.

(6) Is the peace-time use of nuclear energy likely to change our occupations? (Check *one*)
_____ a. No, I don't think so.
_____ b. Yes, but not very much.
_____ c. Yes, a great deal.
_____ d. I don't have any opinion about this.

80. Develop a factual test covering the important points concerning the atom bomb covered in the unit (regarding the fission process, methods of producing atomic materials, developments in nuclear research, contributions of the principal scientists, destruction at Hiroshima and Nagasaki, results of the Bikini test, important dates, how fission is induced, comparative strength derived from nuclear as contrasted with other sources of energy, etc.) and administer this instrument before and after the unit.

81. Develop a factual test covering what has been said and done about the problems growing out of the development of nuclear energy (probable consequences of an atomic war, estimates of likelihood of developing an adequate defense against atom bombs, development of supersonic aircraft, possibilities of sabotage, necessity for rigid inspections, consequent loss of personal freedoms, need for world control of atomic energy, proposals for world control, estimates of time remaining, etc.) and administer this instrument at the beginning and again at the conclusion of the unit.

82. Develop a factual test covering what has been predicted and what has been achieved to date concerning the beneficial effects of nuclear energy if and when it is applied to peace-time uses (medical uses, heat, power, transportation).

Appendix B

Rural Civil Defense Youth Program

> *Elementary and secondary school students in rural America were not immune from the atomics curriculum and related activities to prepare them for an atomic attack. The Office of Civil and Defense Mobilization published* The Rural Civil Defense Youth Program: Leader's Guide *(circa 1960) for educators and community leaders responsible for leading and teaching these boys and girls about civil defense. Here is an excerpt.*

Civil Defense is based on the first law of nature: self-preservation. It is needed and can be had by every family in every home, everywhere. First, and young people share this responsibility, people must learn what the hazards are and what protective steps to take—then take them. This program gives you the facts and helps you to teach the youth you serve.

Introduction

This Rural Civil Defense Program is to help rural boys and girls live and thrive and be better citizens in tomorrow's world by doing their share now in America's civil defense mission. That mission is to save lives and protect property in event of enemy attack, or in a natural disaster such as a tornado or flood.

Communism is the enemy of our democracy. Russia has nuclear weapons which might be used against us, and other hostile nations may soon have them. It is an age of danger—but not hopeless danger. In recent Congressional Hearings, it was brought out that:
- nuclear war would not mean unlimited destruction.
- even if attacked, the United States would be able to recover almost completely in about ten years if we have adequate civil defense.
- in considering a possible enemy attack with 263 nuclear weapons,

The Rural Civil Defense Youth Program

LEADER'S GUIDE

even a moderate shelter program could reduce casualties to approximately one-third of those who would die if there is no protection at all. A more extensive program could reduce the overall fatalities from 25 percent of the population to approximately 3 percent.
- the first requirement is protection against radioactive fallout.
- the program of family fallout shelters depends on simple tools and simple techniques and need not be expensive.
- unreadiness of the American people to survive a nuclear war can greatly undermine our capability to resist Soviet "nuclear blackmail."
- if civil defense protective measures are taken, America's ability to survive would serve to reduce the likelihood of nuclear war.
- we must have civil defense if our Nation is to withstand and recover from nuclear attack.

Objectives

With these facts in mind, you will want to encourage your boys and girls to work on the following objectives:

A. To learn the basic purpose of the Rural Civil Defense Program, one must learn to help teach rural families to protect their lives and property.
  1. To learn what radioactive fallout is; why it is harmful to people, animals, and crops; how to be protected from it; and
  2. To assist their own families in planning, providing, and preparing a family shelter for protection from radioactive fallout, as well as from tornadoes and other disasters; and if they live on a farm, to assist in making a farm protection plan.
B. To help others understand civil defense measures and become capable of caring for themselves in an emergency.
C. To contribute to community survival, plans and actions in either a wartime or a natural disaster.
D. To have a vital part in defending American ideals, to help deter war, and to assist in national survival and recovery if enemy attack should come.

Youth Group Activities

In community service projects and group activities, one of the first things for the boys and girls to decide is what segment of civil defense they want to emphasize.

There are many subjects from which to choose, some of which are suggested, in the following list. After reviewing the subject, the group can decide whether to undertake public information programs, help in community civil defense, encourage other boys and girls to become active in civil defense, or select other activities.
1. Radioactive Fallout: where it comes from; what it is; how it is carried by winds; effects of; how to protect from; decay of radiation; local hazard based on possible targets, wind patterns, etc.
2. Radiological Defense: monitoring, public information, etc.
3. Attack warning; local receipt, public dissemination.
4. CONELRAD: reasons for; what it is, local reception.
5. Family Fallout Shelters:
    a. To use what is available.
    b. To improve.
    c. To make adequate.
    d. Construction details; shielding, entrance, ventilation, etc.
    e. Daily uses for shelter area.
6. Wallet card; its contents and meaning.
7. Survival—Two weeks on your own:
    a. Food, water, clothing, etc.
    b. Family Action Planning.
8. Farm Protection:
    a. Animals, crops, and soil.
    b. Exposure time.
    c. Farm Action Planning
9. Family-Community Civil Defense cooperation; youth assistance.
10. Civil Defense in your community, including assignments of responsibility in County Plan; who is taking what actions; youth assistance.
11. Civil Defense in your county, including responsibilities assigned to county by State Survival Plan; who is taking what actions.
12. What Communism is; what and why it threatens; why civil defense helps deter war and assists in national survival and recovery if we are attacked.

Activities

The following list is suggestive; there are many other worthwhile activities which may be undertaken:
1. Make a tour of a family fallout shelter.
2. Invite the youth representative on the County Action Committee

on Rural Civil Defense (if none, ask that one be appointed from any group working on the Rural Civil Defense Youth Project) to discuss committee activities at one of your meetings.
3. Prepare and give two or more demonstrations at meetings of your own or other groups.
4. Interest a group in another community in Rural Civil Defense and assist them in starting work on the program.

Individual Youth Projects

A. Any of the following personal projects will be beneficial both to youth and to their families. The more boys and girls have learned about civil defense in their group activities, the more they can accomplish individually.
B. Project I—Planning the Family Fallout Shelter
  1. The youth and his family learn about the need for a Family Fallout Shelter; decide what is their best available protection for immediate use; and whether it may be improved or whether new construction will be necessary to provide them fallout protection.
  2. They make decisions and plans on the following points:
    a. Amount of space required (based on number of persons to be sheltered and amount of space needed by each).
    b. Location of family shelter (based on locations available, expense, and other advantages and disadvantages).
    c. Specifications for shelter (shielding material, supports, entrance-way requirements, ventilation, built-in equipment, etc.).
    d. Types of construction to be used (based on materials available, cost, etc.).
    e. Supplies required (for family living for two weeks).
    f. Use of the shelter as a playroom, workshop, den, photographic darkroom, or other purpose.
  3. They make drawings, sketches, and lists to help visualize the above points.
  4. They prepare a Family Action Plan (may be a simple checklist) showing who will be in charge, who next in "command," what actions are to be taken by each member of family on receipt of warning, etc.
C. Project II—Building and Testing Family Fallout Shelter

1. This is the follow-up to Project I, and covers actual construction and preparation of the family fallout shelter, either improvements on existing buildings or a new structure ready to use; plus rehearsal of the family taking shelter according to Family Action Plan and staying in shelter for at least several hours to test the plan.
D. Project III—Farm Protection Planning
  1. The farm youth and his family learn about fall-out on the farm, why it is harmful to animals and crops; how to protect from it; and the youth cooperates in helping his family decide on plans for farm protection.

Decisions for farm protection plans should be based on consideration of all factors. The following list is suggestive:

  a. How many farm animals there are; which are most valuable.
  b. What buildings are available for livestock shelter; which provides most protection and what animals should be put into each building in emergency; how many animals would be without protection, and where they might be confined.
  c. What could be done to make the farm buildings more protective (now; later).
  d. How to provide stock with fresh food and water if family is in shelter; how long stock could go untended; what radiation risks farmer might take to care for stock, and how long he might remain outside shelter.
  e. What could be done to protect stored crops.
  f. Whether any pre-emergency actions should be taken about power, water, fuel, or other farm needs.

Recognition

To all boys and girls, accomplishment of tasks and the joy of participation with others brings its own rewards. Many of the activities outlined in this Guide are concerned with information programs, where the focus of attention is upon the youth who present them; often these projects will be reported in local newspapers. In the home, family, and farm protection activities, boys and girls will take pride in achievement; in knowing they have contributed to personal and national safety; and in knowing their families have set good examples for others to follow.

It will encourage your young people to do more and better civil defense work if you and others express appreciation and satisfaction with their efforts. OCDM Certificates of Achievement are available through your State Civil Defense Director if your organization approves and wants to use them.

Your own best reward for working on this program will be your deep sense of satisfaction that you have helped boys and girls do their share in safeguarding America for their own sakes and for their nation's future.

Appendix C

Glossary of Atomic Terms

During the early years of the atomic age, particularly in the 1950s, after the Soviet Union developed a hydrogen bomb, thus escalating the threat of nuclear war, elementary and secondary teachers and their students were encouraged to develop a fundamental knowledge of atomic and civil defense terms. The following is a compiled glossary of atomic terms from three sources: the New York State Civil Defense Commission, the U.S. Civil Defense Education Project, and the Oregon State Civil Defense Agency.

Active Material. Fissionable material such as U-235 or Pu-239.

Aiming Area. That geographic area within which it is assumed an enemy would most probably place one or more nuclear weapons to assure accomplishment of his mission.

Air Burst. The explosion of an atomic weapon in the air, above land or water, or at a height greater than the maximum radius of the fireball.

Air Raid Warning Signals. The warning signals transmitted from the Divisions of the United States Air Force, through the Federal Civil Defense Administration and state channels, to the local civil defense authorities for dissemination to civil defense personnel and to the public.

Alpha Particle. A helium nucleus, consisting of two protons and two neutrons, with a double positive charge.

Assembly Area or Staging Area. A locality beyond the evacuation perimeter to which designated civil defense equipment and personnel from a target area is dispersed and to which mobile support equipment and personnel is brought forward when conditions in the target area permit their employment.

Atom. The smallest particle of an element which still retains all characteristics of the element.

Atomic Attack. An attack by any of the nuclear weapons, especially the atomic bomb and the hydrogen bomb.

Atomic Bomb. A bomb whose potency is based on the fission or splitting of the atoms of a heavy element such as uranium or plutonium into lighter elements, the sum of whose masses is somewhat less than the mass of the original element. The difference in mass appears in the form of energy. The fission is produced as a result of the bombardment of atomic nuclei by neutrons.

Atomic Energy. All forms of energy released in the course of nuclear fission or nuclear transformation.

Atomic Energy Commission (United States). The official organization in charge of internal control of all phases of nuclear research and atomic energy projects in the United States. It also controls uranium mining, production, and utilization.

Atomic Explosion. Same as nuclear explosion. The detonation of a nuclear weapon. See Nuclear Weapons.

Atomic Mass Unit. One-sixteenth the weight of an oxygen atom expressed in grams, equivalent to 1.66×10^{-24} grams.

Atomic Number. The number of protons in the nucleus, hence the number of positive charges on the nucleus. It is also the number of electrons outside the nucleus of a neutral atom. Symbol: Z.

Atomic Transmutation. The changing of an atom into an atom of different atomic number, that is, into an atom of a different element. The process occurs in nature in the course of the disintegration of the various radioactive elements and may also be affected by artificial means, such as by bombardment with neutrons, alpha particles, gamma radiation, and others.

Atomic Weight. Relative weight of an atom of an element expressed in atomic mass units.

Attack. Attack includes actual or imminent enemy attack in any manner, whether by sabotage or by the use of conventional weapons, nuclear devices, or chemical or biological means.

Attack Warning Devices. Devices, such as horns, sirens, whistles, and special communications facilities, used for warning the public of likely, imminent, or actual attack, and for indicating that there is no further immediate danger.

Attack Warning Officer. An Office of Civil and Defense Mobilization representative located at each of the three Air Defense warning centers (Eastern, Central, Western) who is responsible for disseminating civil defense information over NAWAS (National Attack Warning System).

Attenuation. Reduction in intensity of nuclear radiation, thermal radiation, or air blast by the atmosphere.

Beta Particle. A charged particle emitted from the nucleus and having a mass and charge equal in magnitude to those of the electron.

Binding Energy. A quantity of energy equivalent to the difference in mass between the sum of the masses of component particles in the actual mass of the nucleus.

Biological Warfare. The intentional use of living organisms or their toxic products to cause death, disability, or damage in men, animals, or plants. Also known as bacteriological or germ warfare.

Blast. The pressure effects of the destructive shock wave produced by an explosion. The explosion causes an almost instantaneous expansion of hot gases and, thereby, produces a shock wave which is transmitted outward from the point of the explosion.

Chain Reaction. Any chemical or nuclear process in which some of the products of the process are instrumental in the continuation or magnification of the process.

Civil Defense. Civil defense consists of all the activities and measures aimed at minimizing the effects upon the civilian population caused by enemy attack. Civil defense does not apply to disasters other than those caused by attack.

CONELRAD. Contraction of the words "CONtrol of ELectromagnetic RADiation." An emergency system of radio broadcasting operations which permits standard (AM) radio stations to remain on the air to bring civilian defense information and instructions to the public, but prevents an enemy from using the radio waves as a navigation aid.

Conservation of Mass and Energy. The principle that energy and mass are interchangeable in accordance with the equation $E = mc^2$; where E is energy, m is mass, and c is velocity of light.

Contamination. The deposit of radioactive materials, such as fission fragments or radiological warfare agents, on any object or surface.

Contamination Area. An area containing dangerous amounts of residual radiation.

Control Center. An operations headquarters for the direction and control of civil defense activities during an emergency. The communications nerve center for operations, where attack warnings, damage reports, and other information are received. Every city will have at least one control center. Three functions: to receive information, to issue instructions, and to maintain liaison with higher civil defense authorities.

Crater. The pit, depression, or cavity formed in the surface of the earth by an explosion. This may range from saucer-shaped to conical, depending largely on the depth of the burst; the nearer to the surface the detonation occurs, the shallower the crater.

Critical Mass. For a fissionable material, one critical mass is the minimum amount of a material in the proper environment which will just support a chain reaction.

Critical Size. For a fissionable material, the minimum amount of a material which will support a chain reaction.

Critical Target Area. An area that has been designated as being among the most probable targets in the United States for an enemy attack.

Currie. Rate of radioactive decay, 37 billion disintegrations per second.

Decay. Disintegration of the nucleus of an unstable element by the spontaneous emission of charged particles and/or protons.

Decontamination. Removal of radioactive materials or any foreign materials.

Deuterium. A heavy isotope of hydrogen having one proton and one neutron in the nucleus. Symbol: D or $_1H^2$.

Dismissal Signal. This signal, consisting of a single steady blast for thirty seconds on air raid sirens, horns, or whistles, indicates that the civil defense forces are dismissed.

Dose (Dosage). Amount of radiation delivered to a specified area or volume, or to the whole body.

Dose Rate. The amount of nuclear radiation received by a person per unit time. Sometimes used as a measure of nuclear radiation intensity.

Dosimeter. Instrument used to detect and measure an accumulated dosage of radiation.

Electromagnetic Radiation. Radiation made up of oscillating electric and magnetic fields and propagating with the speed of light. Includes gamma radiation, X-rays, ultraviolet light, visible light, infrared radiation, radar, and radio waves.

Electron. Negatively charged particle which is a constituent of every atom, found external to the nucleus, revolving in an outer orbit. Its mass is 0.00055 atomic mass units, considered as electromagnetic in character. Its diameter is thought to be 10^{-13}cm.

Electron Volt. Amount of energy gained by an electron in passing through a difference of potential of 1 volt. Abbreviated ev.

Element. A substance all of whose atoms have the same atomic number.

Emergency Aid Area. A designated geographical area beyond the evacuation perimeter, adjacent to one or more evacuation routes, in which emergency assistance is provided only to those evacuees who require first aid or fuel for their vehicles to enable them to continue to their destinations and in which evacuating civil defense and medical personnel are initially assembled.

Energy. The capacity for performing work. Potential energy is the energy inherent in a body because of its position with reference to other bodies. Kinetic energy is the energy possessed by a mass because of its motion. CGS units: gm-cm^2/sec^2, or erg.

Evacuation. The organized removal, under civil defense direction and control, of the civilian population or certain sections of the public from areas of actual or threatened destruction or extreme danger to relatively safe areas. (Also see **Strategic Evacuation** and **Tactical Evacuation**.)

Evacuation Area. The area from which people must evacuate for their own protection against the effects (heat, blast, and fallout) of nuclear weapons, when the evacuation signal is sounded. It includes the entire area within the evacuation perimeter.

Evacuation Perimeter. An imaginary line, the outer limits of the evacuation area. For a five-megaton weapon, it lies generally twelve miles from the outer limit of the aiming area. For a one-megaton weapon, it lies generally eight miles from the outer limits of the aiming area.

Fallout. The process of the gradual settling out of particles and the rapid fall of larger particles thrown up by the explosion. The major fallout area may extend from the crater or immediate vicinity of the detonation a distance of many miles, depending upon meteorological and surface conditions. Detectable amounts of fallout may occur over distances of hundreds or thousands of miles for several months after an explosion.

Filter Centers. Installations of the United States Air Force, operated by Air Force personnel assisted by volunteer members of the Aircraft Warning Service, which are set up as strategic central points to receive, plot, and filter the information sent in by the Ground Observer Corps concerning the flight of aircraft, for the purpose of detecting hostile or unidentified plans, and to forward reports to operating units of the United States Air Force.

Fireball. The luminous sphere which begins to form a few millionths of a second after a burst of an atomic bomb.

Fission Products. Elements and/or particles created by fission.

Flash Burns. Those burns caused by instantaneous thermal radiation from the fireball.

Gamma Ray. High frequency electromagnetic radiation with a range of wavelength from 10^{-12} cm., emitted from the nucleus.

Geiger Counter. A low dose rate, high sensitivity detector of beta and measurer of gamma radiation.

Ground Observer Corps. A section of the civil defense Aircraft Warning Service composed of volunteers who man the special ground observation posts and Air Force filter centers for the purpose of detecting and tracing the flight of unidentified or hostile aircraft, and who report such information to operating units of the United States Air Force.

Ground Zero (G.Z.). That point on the earth's surface directly below, at or above where an atomic bomb is detonated.

Half Life. The time required for one-half of the atoms of the substance originally present to decay by radioactivity.

Half Thickness. The thickness of absorbing material necessary to reduce the intensity of radiation by one-half.

Heavy Water. The popular name for water which is composed of two atoms of deuterium and one atom of oxygen.

Hydrogen Atom. The atom of lightest mass and simplest atomic and nuclear structure, consisting of one proton with one orbital electron. Its mass is 1.008123 amu.

Implosion-Type Weapon (or Implosion Weapon). The type of atomic weapon in which a sub critical configuration of fissionable material is compressed into a supercritical state by a centrally directed radial shock to produce an atomic explosion.

Induced Radioactivity. Radioactivity resulting from certain nuclear reactions in which exposure to radiation results in the production of unstable nuclei. Many materials near an atomic explosion enter into this type of reaction, notably as a result of neutron bombardment.

Initial Radiation. The nuclear radiation accompanying an atomic explosion emitted from the resultant ball of fire and atomic cloud (immediate radiation). It includes the neutrons and gamma rays given off at the instant of the explosion and the alpha, beta and gamma rays emitted into the rising ball of fire and column of smoke. In contrast to residual radiation, its effect on persons and objects on the earth's surface is terminated about 90 seconds after the explosion because of the removal of the final source (fission products in the atomic cloud) from within radiation range of the earth at the end of that period of time.

Ion. An atomic particle carrying either a plus or a minus charge, caused by an excess or deficiency of electrons.

Ion Chamber. A container in which ionizations are allowed to take place and in which all the original ions are collected.

Ionization. The production of charged particles (ions) by dislodging electrons from atoms or molecules.

Isotope. One of two or more forms of an element having the same atomic number and hence occupying the same position in the periodic table, but having a different mass number. This is due to a different number of neutrons in the nucleus. There is no difference chemically in isotopes.

Kiloton (KT). A unit of measurement of the yield, or energy release. That amount of energy released from 1,000 tons of TNT (10^{12} calories).

Kinetic Energy. The energy which a body possesses by virtue of its mass and velocity. The equation is: $KE = 1/2\ mv.^2$

Mass Care. The provision of food, clothing, and lodging, immediately following attack, to all persons requiring such assistance, without regard to their ability to pay.

Mass Number. The total number of protons and neutrons in the nucleus.

Megaton (MT). A unit of measurement of the yield, or energy released. That amount of energy released from one million tons of TNT (10^{15} calories).

MEV (Mev.). The symbol of one million electron volts.

Mobilization Roads. Roads and streets within a county or city, which are designated by the local civil defense director for the use of civil defense forces in combatting a disaster in the area, and which should be completely policed. Police assigned to traffic control on local mobilization roads will be under the command of the local civil defense director.

Molecule. The ultimate unit quantity of a compound which can exist by itself and retain all the properties of an original substance.

Monitoring. Locating radioactive contamination by means of survey instruments which indicate the residual radioactivity in terms of radiation intensity.

MPE. Maximum permissible exposure.

NAWAS (National Attack Warning System). The federal government's portion of the nationwide attack warning system utilized for dissemination of warning information from the Office of Civil Defense and Mobilization warning centers to civil defense warning point locations in each state.

Neutron. An elementary nuclear particle or mass number 1. It is believed to be a constituent particle of all nuclei of a mass number greater than 1. Its rest mass is 1.00894 amu.

Nominal Bomb. An atomic bomb with a destructive force equivalent to 20,000 tons (20 kilotons) of TNT. Atomic bombs of this destructive magnitude were used in the attacks on Hiroshima and Nagasaki, Japan.

Nuclear Fission. A special type of nuclear transformation characterized by the splitting of a nucleus into at least two other nuclei and the release of a relatively large amount of energy.

Nuclear Fusion. The act of coalescing two or more nuclei.

Nuclear Radiation. Any or all of the radiations emitted as a result of a nuclear transformation. The radiations include gamma radiation (of electromagnetic character) and particle radiations (alpha particles, positive and negative beta particles and neutrons).

Nuclear Weapons. Weapons of warfare which employ nuclear energy to produce explosions of great destructive force, such as the atomic bomb, hydrogen bomb, and atomic artillery shell.

Nucleon. Common name applied to any particle found within the nucleus.

Nucleus. The massive, hard core of the atom whose diameter is about 10^{-12} to 10^{-13} cm.

Plutonium (Pu^{239}). A chemical element produced artificially by the capture of neutrons in the uranium 238 nucleus. Quantity production of plutonium is accomplished in nuclear reactors called "piles." Pu^{239} is one of the fissionable materials used in making atomic weapons.

Proton. A nuclear particle with a positive electric charge equal numerically to the charge of the electron and having a mass of 1.007575 amu.

Radiation. A method of transmission of energy. Specifically:
(1) Any electromagnetic wave (quantum).
(2) Any moving electron or nuclear particle, charged or uncharged, emitted by a radioactive substance.

Radioactivity. The process whereby certain nuclides undergo spontaneous atomic disintegration in which energy is liberated, generally resulting in the formation of new nuclides. The process is accompanied by the emission of one or more types of radiation, such as alpha particles, beta particles and gamma radiation.

Reaction. Any process involving a chemical or nuclear change.

Residual Contamination. That nuclear radiation that is left after the detonation. Usually refers to radiation from the cloud or from fission products deposited on the ground.

Residual Radiation. Nuclear radiation emitted by the radioactive material deposited after an atomic burst or an attack with radiological warfare agents. Following an atomic burst, the radioactive residue is in the form of fission products, unfissioned nuclear material and material (such as earth and water constituents and exposed equipment) in which radioactivity may have been induced by neutron bombardment. It is sometimes referred to as lingering radiation.

Roentgen. The quantity of X- or gamma-radiation which produces 1 esu of positive or negative electricity/cm.³ of air at standard temperature and pressure or 2.083×10^9 ion pairs/cm.³ of dry air.

Roentgen Equivalent Man (or Mammal) (rem). The quantity of ionizing radiation of any type which, when absorbed by man (or other mammal), produces a physiological effect equivalent to that produced by the absorption of one roentgen of X-ray or gamma radiation.

Scintillation Chamber. A type of instrument detector which contains a material in crystalline form which gives off a characteristic light when exposed to nuclear radiation.

Strategic Evacuation. That which is accomplished during a period of international tension indicating a possible attack, when it may be desirable to move certain dependent and/or nonproductive people along with certain industries and services away from danger areas.

Surface Burst. An explosion at the surface of land or water or at a height above the surface less than the maximum fireball radius.

Tactical Evacuation. That which is accomplished during an air raid warning period, when it has been decided that time permits the mass evacuation of people and dispersal of mobile industry and services from target areas.

Target Area. An area designated as a potential target for enemy attack because of its strategic importance as an industry and/or population center.

Thermal Radiation. Radiation emitted as a result of high temperature.

Thermonuclear. An adjective referring to the process involving the fusion of light nuclei such as those of deuterium and tritium.

Underground Burst. An explosion with its center of detonation beneath the surface of the ground.

Underwater Burst. An explosion with its center of detonation beneath the surface of the water.

Uranium. The heaviest naturally occurring element. It consists of three naturally occurring isotopes, U^{238} and U^{235} and U^{234}. U^{235} is a fissionable material and is used in the atomic weapon as a nuclear explosive.

Velocity of Light. 3×10^{10} cm./sec.

Warning Red. Audible signal warning the public and civil defense personnel that enemy attack is imminent and that shelter or cover should be taken as quickly as possible. It is a signal of three minutes duration consisting of either a fluctuating or warbling sound of varying pitch on air raid sirens, or a succession of intermittent blasts on horns or whistles.

Warning White. Signal indicating that the danger of probable or imminent attack is over, or when, after actual attack, the danger of repeated attack is neither probable nor imminent. If the Warning White follows a Warning Red, the Warning White (All Clear) is transmitted to the public as a series of three steady one-minute blasts on air raid sirens, horns, or whistles with periods of two minutes of silence between each blast. If the Warning White follows a Warning Yellow, without an intervening Warning Red, no audible warning signal will be sounded. In this case, the Warning White is a confidential signal transmitted in the same manner as and to the same persons who received the Warning Yellow.

Warning Yellow. A confidential signal indicating that enemy attack is probable. This warning is transmitted to key civil defense personnel and to appropriate school, hospital, and welfare authorities, in order to enable them to carry out their prearranged plans for emergency operations.

Wavelength. The linear distance between any two similar points of two consecutive waves.

Weight. The force with which a body is attracted toward the earth. CGS units: gm.-cm. sec^2.

Yield. The energy release of an atomic weapon, usually expressed in kilotons (1,000 tons) of TNT equivalent.

Chapter Notes

Preface

1. Robert A. Jacobs, "Atomic Kids: Duck and Cover and Atomic Alert Teach American Children How to Survive Atomic Attack," *Film & History* 40, no. 1 (Spring 2010), 25.
2. See Dee Garrison, *Bracing for Armageddon: Why Civil Defense Never Worked* (New York: Oxford University Press, 2006), 19–20.
3. See John F. Kennedy Presidential Library and Museum's audio recordings.

Introduction

1. *Civil Defense and the Schools* (New York: New York State Civil Defense Commission, 1953), vii.
2. "Duck and Cover: The Children Look at Atom-Raid Drills," *Children and the Threat of Nuclear War* (New York: The Child Study Association of America, 1964), 34–45.
3. JoAnne Brown, "A Is for Atom, B Is for Bomb," *The Journal of American History* 75 (June 1988): 74–81.
4. See Florence Gelbond, "The Impact of the Atomic Bomb on Education," *The Social Studies* 65 (March 1974): 110.
5. Willem J. Van Der Grinten, "Not Only Science Teachers," *The Clearing House*, 24, no. 8 (April 1950): 487.
6. Sibylle Escalona, "Children and the Threat of Nuclear War," *Children and the Threat of Nuclear War* (New York: The Child Study Association of America, 1964), 8.
7. Quote from Joseph C. Goulden, *The Best Years 1945–1950* (New York: Antheneum, 1976), 428.
8. Escalona, "Children and the Threat of Nuclear War," 24.
9. Norman Cousins, *Modern Man Is Obsolete* (New York: Viking, 1945), 7.
10. John Hersey, *Hiroshima* (New York: Alfred K. Knopf, 1946).
11. A. M. Holladay, "The Atom and Educator," *Peabody Journal of Education*, 25, 2 (September 1947): 103.
12. Paul Boyer, *By the Bomb's Early Light* (New York: Pantheon, 1985), 15.
13. Hersey, *Hiroshima*, 116.
14. Boyer, *By the Bomb's Early Light*, 21.
15. Margaret Gowing, *Britain and Atomic Energy 1939–1945* (New York: St. Martin's, 1964), 386.
16. David Lilienthal, "Atomic Energy … and You," *Senior Scholastic*, 12 April 1948, 3.
17. Edgar Dale, "It's a Nice World—Wasn't It?" *The High School Journal* 29 (October 1946): 176–180.
18. C.S. Kazdan, "Postwar Problems in Education," *Journal of Educational Sociology* 19, no. 6 (February 1946): 352.
19. Kazdan, "Postwar Problems in Education," 357–358.
20. See Andrew D. Grossman, *Neither Dead Nor Red* (New York: Routledge, 2001), 39.
21. *Interim Civil Defense Instructions for Schools and Colleges* (Washington, D.C.: Government Printing Office, 1951), 5. Also see Andrew Grossman, *Neither Dead Nor Red*, 158.
22. Clara P. McMahon, "Civil Defense and Education Goals," *The Elementary School Journal* 53 (April 1953): 440–442.

23. See William Graebner, *The Age of Doubt: American Thought and Culture in the 1940s* (Boston: Twyane, 1991).
24. Kristina Zarlengo, "Civilian Threat, the Suburban Citadel, and Atomic Age Women," *Signs* 24 (Summer 1999): 940.
25. Dale, "It's a Nice World—Wasn't It?" 180.
26. *Nuclear Survival: A Resource Handbook* (Albany: University of the State of New York, 1961), 3.

Chapter 1

1. Arthur Compton, "Introduction," *One World or None* (New York: McGraw-Hill, 1946), V.
2. Andrew Hartman, *Education and the Cold War: The Battle for the American School* (New York: Palgrave McMillan, 2008), 55.
3. Hartmann, *Education and the Cold War,* 56.
4. "'Two to Five Years,'" *The Journal of Education* 128 (1945): 297.
5. Harold Urey, "How Does It All Add Up?" *One World or None*, 58.
6. Lewis Todd, "Atomic Energy and the Coming Revolution in Education," *School and Society* 62 (1945): 251–257.
7. W.H. McFarland, "World Unity in the Classroom," *The Journal of Education* 129 (March 1946): 97.
8. McFarland, "World Unity in the Classroom," 96.
9. "Teaching International Understanding," *The Phi Delta Kappan* 28, no. 2 (October 1946), 91–92.
10. "Teaching International Understanding," 91–95.
11. Sam Rayburn, "That Civilization May Survive," *School Life* 28 (October 1945): 9–10.
12. Tracy Mygatt, "World Government Is Common Sense," *Progressive Education* 24 (October 1946): 10–11.
13. Lyle Ashby, "Operation Crossroads," *Journal of the National Education Association* 35 (1946): 292.
14. Robert Hutchins, "The Issues in Education: 1946," *The Educational Record* 27 (1946): 365–375.
15. Harold C. Hand, "Education for Survival," *Educational Leadership* 4 (October 1946), 10.
16. Hand, "Education for Survival," 8.
17. Hand, "Education for Survival," 4–5.
18. Hand, "Education for Survival," 7.
19. Nathaniel Peffer, "Politics Is Peace," *The American Scholar* 15 (1946): 160–166; Sumner Welles, "One Practical Chance—the UNO," *The American Scholar* 15 (1946): 141–143.
20. Herbert Abraham, "A World Organization for Peace," *School Life* 28 (October 1945): 3–4.
21. Merril Bush, "World Organization or Atomic Destruction?," *School and Society,* 23 (November 1946), 353–355.
22. Bush, "World Organization or Atomic Destruction?" 355.
23. Louis Ridenour, "Science and Secrecy," *The American Scholar* 15 (1946): 147–153.
24. Erich Kahler, "The Reality of Utopia," *The American Scholar* 15 (1946): 167.
25. Kahler, "The Reality of Utopia," 179.
26. Alonzo May, "Atomic Energy and the Liberal Arts," *School and Society,* 24 (August 1946), 131–133.
27. Henry Christ, "The Atom Bomb Shakes the Classroom," *Journal of the National Education Association* 35 (1946): 296.
28. John Starie, "Schools and the Atom," *Education* 66 (1946): 501–502.
29. "Psychologists Advise on the Atomic-Bomb Peril," *School and Society*, 8 June 1946, 405–406.
30. Quincy Wright, "Barriers to World Peace," *School Review* 54 (1946): 576–583; "One World and the Teaching of History," *School and Society*, 23 August 1947, 132; John S. Perkins, "Where Is the Social Sciences' Atomic Bomb?," *School and Society*, 17 November 1945, 315–317.
31. "Editorial," *Progressive Education* 24, no. 1 (1946): 16.
32. Edmund Day, "Educational Mobilization in a Free Society," *The Educational Forum* 11 (November 1946): 10; "Pupils Advertise American Education Week," *The Manual Craftsman*, Kansas City (November 1, 1946): 1; "'Atomic Power, Peace' Discussed by R. Riley," *The Southeast Tower*, Kansas City (November 22, 1946): 2; "Speech on Atomic Age," *The Manual Craftsman*, Kansas City (November 27, 1946): 1.

33. John W. Studebaker, "Secondary Education for A New World," *School Life* 29 (October 1946): 3–8.
34. John W. Studebaker, "The High Schools of the Future," *School Life* 29 (April 1947): 306.
35. William Carr, "On the Waging of Peace," *Journal of the National Education Association* 36 (1947): 495–500.
36. Carr, "On the Waging of Peace," 496.
37. Carr, "On the Waging of Peace," 496.
38. Ronald Lora, "Education: Schools as Crucible in Cold War America," *Reshaping America*, ed. Robert Bremner and Gary Reichard (Columbus: Ohio State University, 1982), 237; R. Will Burnett, "The Teacher and Atomic Energy," *Education* 68 (1948): 545.
39. Harry Gail, "Atomic Energy and Education," *Progressive Education* 24, no. 4 (1947): 116–199+.
40. Harold Rugg, "Progressive Education—Which Way?" *Progressive Education* 25, no. 4 (1948): 35–37, 45.
41. Willard Goslin, "A Task for Administrators," *School Life* Supplement 31 (March 1949): 1.
42. David Lilienthal, "Education's Responsibilities," *School Life* Supplement 31 (March 1949): 1–2.
43. Mabel Studebaker, "The Teacher and the Atom," *School Life* Supplement 31 (March 1949): 1.
44. R. Will Burnett, Ryland Crary, and Hubert Evans, "The Minds of Men," *School Life* 31 (March 1949): 11–13.
45. A.H. Lindsey, "I Taught Atomic Energy: With Statements by Thirteen Members of the Class," *Education* 71 (1951): 451–469.
46. For a discussion of the shift toward apathy and atomic anxieties, see Paul Boyer, *By the Bomb's Early Light: American Thought and Culture at the Dawn of the Atomic Age* (New York: Pantheon, 1985), 281–297.
47. Lyman Graybeal, "Democratic Education in a Time of Crisis," *School Activities* 20 (1949): 211–212, 215, 218.
48. John Brooks, "Ends and Means in Teaching a World Order," *Progressive Education* 30, no. 3 (1953): 72–74.
49. R. Will Burnett and Harold Hand, "Educational Implications of the Atomic Age, *Education* 71, 7 (1951): 429–445.
50. Burnett and Hand, "Educational Implications of the Atomic Age," 435.
51. Burnett and Hand, "Educational Implications of the Atomic Age," 439.
52. Burnett and Hand, "Educational Implications of the Atomic Age," 442.

Chapter 2

1. State of New Jersey Division of Civil Defense, *Civil Defense and the School Principal* (Trenton, Division of Civil Defense, 1952): 9.
2. "The Atomic Age," *Life* 20 August 1945, 32.
3. "Adjustment," *New Yorker*, August 18, 1945, 17.
4. "Birth of an Era," *Time* 13 August 1945, 17–18.
5. "The Significant Year," *The* (Kansas City) *East Echo*, 5 September 1945, 1.
6. Aaron Goff, "The Atom and Civilization: Ten Urgent Classroom Duties for Teachers," *The Clearing House* 21, no. 8 (April 1947): 457.
7. Goff, "The Atom and Civilization: Ten Urgent Classroom Duties for Teachers," 460.
8. Goff, "The Atom and Civilization: Ten Urgent Classroom Duties for Teachers," 458.
9. Fletcher G. Watson, "Workshop on Atomic Energy: New England Project for Secondary Schools," *The Clearing House* 23, no. 8 (April 1949): 456–467.
10. Leo Weitz, "A Social Studies Unit on Atomic Energy," *High Points* 31 (February 1949): 14–26.
11. Weitz, "A Social Studies Unit on Atomic Energy," 18.
12. For more information about Oak Ridge, Tennessee, see Denise Kiernan, *The Girls of Atomic City* (New York: Simon & Schuster, 2013).
13. Norman Cousins, *Modern Man Is Obsolete* (New York: Viking, 1945), 7, 23.
14. W. Ogden, "Ridge Kids Use the Atom," *New York Times Magazine* 2 June 1946, 24–25; W.C. Seyfert, "Youth in the Atomic Age," *School Review* 54 (June 1946): 319–320; Philip Kennedy, "Oak Ridge and the Educational Crossroads,"

National Association of Secondary-School Principals Bulletin 30 (October 1946): 81–83.

15. Sally Cartwright, "Where the Atom Bomb Was Born," *Progressive Education* 24 (October 1946): 4; Jeanette E. Sawyer, "Science at Oak Ridge High School," *The Clearing House* 21, no. 6 (February 1947): 361–362; Virginia Hardin Stearns, "Denver's International Relations Club," *The Journal of Education* 133 (May 1950): 138–140.

16. Benjamin Fine, "'A Better World' Courses in New York City Schools," *The Education Digest* 12 (September 1946): 10–11 (reprint from *New York Times*, 19 May 1949).

17. Harold C. Hand, ed., "Living in the Atomic Age: A Resource Unit for Teachers in Secondary Schools," *University of Illinois Bulletin* 23 (December 3, 1946), 1–60.

18. William E. Kane, "An Atomic Age Week," *The School Review* 56, no. 5 (May 1948): 294–298.

19. Harold H. Metcalf, "How Teach the United Nations?," *The Phi Delta Kappan* 32, no. 4 (December 1950): 131.

20. Metcalf, "How Teach the United Nations?" 132.

21. Metcalf, "How Teach the United Nations?" 132.

22. Millard Harmon, "Teaching Science in the Elementary School," *The Elementary School Journal* 50, no. 5 (January 1950): 273–276.

23. Audrey Lindsey, "I Taught Atomic Energy: With Statements By Thirteen Members of the Class," *Education* 71, no. 7 (1951): 451–469.

24. Lindsey, "I Taught Atomic Energy," 465–469.

25. *Atomic Energy and You*, Los Angeles School District, 1953; Rosalie Kirshen, "A Unit on Atomic Energy in the Experience Curriculum," *Education* 71, no. 7 (1951): 451–469; David Hilton and Mary Jeffries, "Atomic Energy in the Classroom and Community," *Journal of Education* 131 (March 1948): 88–89; Hubert Evans and Ryland Crary, "Atomic Education: A Continuing Challenge," *Teachers College Record* 50 (1949): 515–520; Milton J. Gold, "26 High Schools Use Radioisotopes: From Workshop Activity to Radioactivity," *The Clearing House* 28, no. 6 (February 1954): 359–362.

26. Gerrit Zwart, "How a Small High School Meets the Challenge of the Atomic Age: Suffern High School Atomic Energy Club," *School Life* Supplement 35 (September 1953): 147+.

27. "Mouse Traps for Chain Reaction," *School Life* 32 (November 1949): 21–22.

28. Bryan F. Swan and Generose Dunn, "A Unit on Atomic Energy for Junior High School," *The School Review* 62 (April 1954): 231–236.

29. William Reaves, "Organized Extra-Curricular Activities in the High School," *The High School Journal* 34 (May 1951): 130.

30. Stuart Little, "The Friendship Train: Citizenship and Postwar Culture, 1946–1949," *American Studies* 34 (Spring 1993): 35–67; "Hundreds Come to See Friendship Train Loaded With Sacks of Grain for Europe," *The* (Kansas City) *Paseo Press* 5 December 1947, 1; "Jam Session," *Senior Scholastic* 29 September 1948, 35.

31. Boyer, *By the Bomb's Early Life*, 296–297; Richard Robin, "Power and the Atom," *The Journal of Educational Sociology* 22 (January 1949): 350–352; Richard Hitchcock, "Westinghouse Theater of Atoms," *The Journal of Educational Sociology* 22 (January 1949): 353–355; Lillian Wald Key, "Public Opinion and the Atom," *The Journal of Educational Psychology* 22 (January 1949): 356–362; Louis Heil and Joe Musial, "'Splitting the Atom'—Starring Dagwood and Blondie: How It Developed," *The Journal of Educational Psychology* 22 (January 1949): 331–336; "Atomic Energy Book Exhibit for New York's Golden Jubilee," *Publishers Weekly*, 10 July 1948, 140; "Atomic Energy Book Exhibit Touring the Country," *Publishers Weekly*, 27 November 1948, 204.

32. Kenton Clymer, "The Ground Observer Corps," *Journal of Cold War Studies* 15, no. 1 (Winter 2013), 34–52. For President Truman's statement, see the Public Papers of the Presidents, Harry S. Truman Library & Museum.

33. *The Alert America Convoys* (Valley Forge Foundation, 1952), 5.

34. Helen Heffernan, "The School Curriculum in American Education," in

Edgar Fuller and Jim Pearson, eds., *Education in the States: Nationwide Development Since 1900* (Washington, D.C.: National Education Association, 1969), 215–285; "More Scripts for High School Radio Workshops," *Senior Scholastic*, Teacher Edition, 26 September 1951, 27T; *Senior Scholastic*, Teacher Edition, 7 March 1951, 23T; Elizabeth Drake and Lillian Carmen, "A Broadcast for Brotherhood," *School Activities* 21 (October 1949): 56–58, 68; "Six Pupils Give Radio Talks On Europe's Food Problem," *The* (Kansas City) *Westport Crier* 24 September 1947, 1; "TV Show Spotlights Teenagers; Larry Ray Acts as Moderator," *The* (Kansas City) *Southwest Trail*, 25 February 1954, 1; Martha Gable, "Philadelphia Classroom Television," *The Journal of Education* 134 (February 1951), 50–52.

35. *Civil Defense and the Schools* (New York State Civil Defense Commission, 1953), vii.

36. *Civil Defense and the Schools*, 49–56.

37. Earl Peckham, "The Place of Civil Defense in Education," *School and Society* 9 (August 1952), 87–90.

38. Peckham, "The Place of Civil Defense in Education," 90.

Chapter 3

1. New York Civil Defense Commission, *Civil Defense and the Schools* (1953): 35–36.

2. Michael Carey, "The Schools and Civil Defense: The Fifties Revisited," *Teachers College Record* 84, no. 1 (1982), 15–127.

3. Alan Winkler, *Life Under a Cloud* (New York: Oxford University Press, 1993), 4.

4. Michael Yavanditti, "The American People and the Use of Atomic Bombs on Japan: The 1940s," *Historian* 36 (February 1974): 224–226; "Fortune Survey: Use of Atomic Bomb," *Fortune* 33 (December 1945): 305–306+.

5. George Hopkins, "Bombing and the American Conscience During World War II," *Historian* 28 (May 1966): 472–473.

6. Peter Hales, "The Atomic Sublime," *American Studies* 32 (Spring 1991): 5–31.

7. Paul Boyer, *By the Bomb's Early Light: American Thought and Culture at the Dawn of the Atomic Age* (New York: Pantheon, 1985), 21, 32, 281–297.

8. William Graebner, *The Age of Doubt: American Thought and Culture in the 1940s* (Boston: Twyane, 1991), 18, 102, 126, 147.

9. Joseph Lantagne, "Health Interests of 10,000 Secondary School Students," *The Research Quarterly* 23 (October 1952): 330–346.

10. Andrew D. Grossman, *Neither Dead Nor Red* (New York: Routledge, 2001), 44–45.

11. Grossman, *Neither Dead Nor Red*, 84.

12. Urban Fleege, "The Teacher and Civil Defense," *Journal of the National Education Association* 40 (November 1951): 542–543.

13. Fleege, "The Teacher and Civil Defense," 542–543.

14. JoAnne Brown, "'A Is for Atom, B Is for Bomb': Civil Defense in American Public Education, 1948–1963," *The Journal of American History* 75 (June 1988): 76.

15. Bonaro Overstreet, "Understanding Our Fears," *Journal of the National Education Association* 41 (February 1952): 85–86.

16. *Civil Defense and the Schools*, New York State Civil Defense Commission (1953): 44.

17. Val Peterson, "Panic, the Ultimate Weapon?" *Collier's*, 21 August 1953, 99–105.

18. Laurence Sears, "Anxiety in the United States of America," in H. Gordon Hullfish, ed., *Educational Freedom in an Age of Anxiety: Yearbook of the John Dewey Society 1953* (New York: Harper & Brothers, 1953): 26.

19. Sears, "Anxiety in the United States of America," 1.

20. H. Gordon Hullfish, "Education in an Age of Anxiety," in H. Gordon Hullfish, ed., *Educational Freedom in an Age of Anxiety: Yearbook of the John Dewey Society 1953* (New York: Harper & Brothers, 1953): 208.

21. Clara P. McMahon, "Civil Defense and Educational Goals," *The Elementary School Journal* 53 (April 1953): 442.

22. Kimball Wiles and Woodrow

Sugg, "Factors Influencing Curriculum Development, *Review of Education Research* 24, no. 3 (June 1954): 194.

23. "Atomic Energy Here to Stay," *School Life* Supplement 31 (March 1949): 2.

24. *Senior Scholastic*, Teacher Edition, 12 Nov. 1945, 1T. Emphasis in original text.

25. *Senior Scholastic*, Teacher Edition, 2 March 1949, 15T. The magazine told of one high school ordering 8,000 books from the Book Service in one semester, with the average English student reading four to five titles.

26. *Senior Scholastic*, Teacher Edition, 9 December 1953, 1T-2T.

27. *Senior Scholastic*, Teacher Edition, 14 April 1954, 9T.

28. "Psychologists Advise on the Atomic-Bomb Peril," *School and Society*, 8 June 1946: 405–406.

29. Dorothy McClure, "Social-Studies Textbooks and Atomic Energy," *The School Review* 57 (December 1949): 540–546.

30. Hubert Evans, Ryland Crary and C. Glen Hass, *Operation Atomic Vision* (1948: National Association of Secondary-School Principals, Washington, D.C.); "What Is Operation Atomic Vision?," *National Association of Secondary-School Principals Bulletin* 32 (April 1948): 198–204; Evans, Crary and Hass, "Operation Atomic Vision," *Journal of the National Education Association* 37 (October 1948): 439–442; Paul Boyer, *By the Bomb's Early Light: American Thought and Culture at the Dawn of the Atomic Age* (New York: Pantheon Books, 1985), 281–297.

31. *Operation Atomic Vision*, 5

32. *Operation Atomic Vision*, 25, 37, 40.

33. Evans, Crary, and Hass, "Operation Atomic Vision," 442. Emphasis in original text.

34. Evans, Crary, and Hass, "Operation Atomic Vision," 440.

35. Hubert Evans and Ryland Crary, "Atomic Education: A Continuing Challenge," *Teachers College Record* 50 (1949): 519.

36. Benjamin Starr and Abraham Leavitt, "Social Studies and 'Operation Atomic Vision,'" *High Points* 31 (April 1949): 22–32.

37. Kirshen, "A Unit on Atomic Energy in the Experience Curriculum," 27–31.

38. Kenneth D. Rose, *One Nation Underground: The Fallout Shelter in American Culture* (New York: New York University Press, 2001), 129–130. Also see Grossman, *Neither Dead Nor Red*, 84.

39. George L. Glasheen, "What Schools Are Doing in Atomic Energy Education," *School Life* 35 (September 1953): 152.

40. L. J. Mauth, "Prevention of Panic in Elementary-School Children," *The Journal of Education* 137, no. 2 (November 1954): 10–14.

41. S. Mary Amatora, "Emotional Stability of Children in the Atomic Age," *Education* 71, no. 7 (1951): 446–450.

42. Mauth, "Prevention of Panic in Elementary-School Children," 14.

43. *Civil Defense Manual*, Georgia Schools (September 1952): 21–27.

44. *Civil Defense for Schools* (Pennsylvania) State Council of Civil Defense (1952): 8.

45. *Civil Defense for Schools*, 12.

46. *Civil Defense for Schools*, 19.

47. *Civil Defense in Oregon Schools*, introduction.

48. *Civil Defense in Oregon Schools*, 14.

49. *Civil Defense in Oregon Schools*, 39.

50. David Jenkins, "Social Engineering in Educational Change: An Outline of Method," *Progressive Education* 26, no. 7 (1949): 193–197; Kenneth Benne, "Democratic Ethics in Social Engineering," *Progressive Education* 26, no. 7 (1949): 201–207; Ralph White, "Ultimate and New: Ultimate Democratic Values, *Progressive Education* 27, no. 6 (1950): 165–171. For more discussion of the social engineering of postwar adolescents, see William Graebner, "The 'Containment' of Juvenile Delinquency: Social Engineering and American Youth Culture in the Postwar Era," *American Studies* 27 (Spring 1986): 81–97; and Graebner, *Coming of Age in Buffalo: Youth and Authority in the Postwar Era* (Philadelphia: Temple University Press, 1990).

Chapter 4

1. Federal Civil Defense Administration, *Civil Defense in Schools* (Washington,

D.C.: U.S. Government Printing Office, 1952), 2.

2. Public Papers of the Presidents, Harry S. Truman Library (Independence, Missouri).

3. Richard Noyes, "The Teacher and the Atom Bomb," *Journal of the National Education Association* 35, no. 6 (1946): 296–297.

4. See Willis Rudy, *Schools in the Age of Mass Culture* (Englewood, N.J.: Prentice Hall, 1965); Andrew Hartmann, *Education and the Cold War: The Battle for the American School* (New York: Palgrave McMillan, 2008). Among the numerous books published during this period dealing with the state of education, including life adjustment education, are C. Wayne Gordon, *The Social System of the High School* (New York: Free Press of Glencoe, 1957), Henry Ehlers, ed., *Crucial Issues in Education: An Anthology* (New York: Holt, 1955), and James Bryant Conant, *Education and Liberty: The Role of the Schools in a Modern Democracy* (Cambridge: Harvard University Press, 1953), Harl Douglass, *Education for Life Adjustment: Its Meaning and Implication* (New York: Ronald Press, 1950), and Franklin Zeran, ed., *Life Adjustment Education in Action* (New York: Chartwell House, 1953). For additional insight into the history of progressive or life-adjustment education in the 1940s and 1950s, see Joel Spring, *The Sorting Machine: National Educational Policy Since 1945* (New York: David McKay, 1976), Lawrence Cremin, *Transformation of the School* (New York: Alfred A. Knopf, 1961) *American Education, The Metropolitan Experience, 1876–1980* (New York: Harper & Row, 1988), and Dorothy Elizabeth Broder, "Life Adjustment Education: An Historical Study of a Program of the United States Office of Education, 1945–1954," Dissertation, Teachers College, Columbia University, 1976.

5. Adolph Unruh, "Life Adjustment Education—A Definition," *Progressive Education* 29, no. 4 (1952): 137–141.

6. Chester Diettert, "To Keep Democracy at Its Best," *School Activities* 21 (March 1950): 213.

7. Cremin, *Transformation of the School*, 348.

8. For a discussion about the criticisms of life-adjustment education, see Rudy, *Schools in the Age of Mass Culture*, 324; and Cremin, *Transformation of the School*, 338.

9. Rudy, *Schools in the Age of Mass Culture*, 324.

10. Samuel Capen, "The Truth Will Prevail," *School and Society*, 9 September 1947, 178; Curtis MacDougall, "Language and Human Welfare," *Progressive Education* 25, no. 1 (1947): 269.

11. John W. Studebaker, *Education and the Fate of Democracy* (Los Angeles: University of California Press, 1948).

12. George Gallup, *The Gallup Poll: Public Opinion 1935–1971* (New York: Random House, 1972): 675, 916.

13. Edwin Miner, "National Conference on Zeal for American Democracy," *School Life* 30 (May 1948): 3–5; Ward Keesecker, "Duty of Teachers to Promote Ideals and Principles of American Democracy," *School Life* 30 (February 1948): 31–33.

14. Charles Peters, *Teaching High School History and Social Studies for Citizenship Training* (Coral Gables: University of Miami Press, 1948), 38–40.

15. Ron Davis, "Laboratory Practice in Protective Skills," *School Life* 35 (1953): 158–159; "Another Citizenship Program," *The Journal of Education* 133 (May 1950): 135; "For Citizenship and Moral Training," *Senior Scholastic*, 7 November 1951: 12–13.

16. "Dix's Students Discuss Current World Topics," *The East* (Kansas City) *Echo* 13 March 1952, 1; Hazel Torrens, "Current Events in the Ninth Grade," *The Education Digest* 12 (December 1946): 22–23; Barbara York, "Quincy High School's P.D. Course," *The Journal of Education* 131 (September 1948): 218–219; "Kansas Will Train Youth As Citizens," *The Journal of Education* 133 (September 1950): 189; *Senior Scholastic*, Teacher Edition, 15 April 1946, 5T.

17. W.H. McFarland, "World Unity in the Classroom," *The Journal of Education* 129 (March 1946): 97.

18. John W. Studebaker, "Communism's Challenge to American Education," *School Life* 30 (February 1948): 1–7.

19. David Lilienthal, "Democracy and the Atom," *Progressive Education* 25, no. 3 (1948): 2–5+.

20. Kendric Marshall, "Teachers and the International Crisis," *School Life* 30 (June 1948): 2; Mattie Pinette, "School and Community Face the Atomic Age," *School Life* 35 (September 1953): 155.

21. Ruth Strong, "Students Interpret the School to the Community," *The High School Journal* 34 (March 1951): 66–68.

22. "Public Schools Urged to Stress Citizenship and United Nations," *The Journal of Education* 133 (November 1950): 239.

23. J.G. Umstattd, "Contributions of the Secondary Schools in the Present World Situation," *The High School Journal* 34 (May 1951): 145–152.

24. Sumner Pike, "The Promise of Atomic Energy," *Education* 71, no. 7 (1951): 407–413; Warren Austin, "Atomic Weapons and World Peace," *Education* 71, no. 7 (1951): 414–419.

25. Homer Higbee, "The Social, Economic, and Political Implications of Atomic Energy," *Education* 71, no. 7 (1951): 420–428.

26. Higbee, "The Social, Economic, and Political Implications of Atomic Energy," 420, 422, 423.

27. Higbee, "The Social, Economic, and Political Implications of Atomic Energy," 425, 427.

28. R.J. Blakely, "Living Without Fear in a Century of Continuing Crisis," *School Life* 35 (September 1953): 145–146.

29. Ryland Crary, "Curriculum Adaptation to Changing Needs," *School Life* 35 (September 1953): 157, 160.

30. "Education and National Security," *Journal of the National Education Association* 41 (January 1952): 21–22.

31. J. Clyde Johnson, "Teaching Democratic Skills and Attitudes," *The High School Journal* 35 (February 1952): 137–142.

32. Arnold Perry, "Fundamental Education and the Defense of Democracy," *The High School Journal* 38 (January 1955): 120.

33. Howard Hightower, "On War and Peace," *Progressive Education* 29, no. 7 (1952): 253.

34. Earl James McGrath, *Education: The Wellspring of Democracy* (Tuscaloosa: University of Alabama Press, 1951), 70, 127.

35. Louis Kaplan, "The Need for Creative Education," *The Journal of Education* 131 (November 1948): 241–242.

36. Kaplan, "The Need for Creative Education," 242.

Chapter 5

1. Committee on Fallout Protection, *Survival in a Nuclear Attack: Plan for Protection from Radioactive Fallout* (New York: New York State Civil Defense Commission, 1960): 4.

2. Peter Rose, "The Public and the Threat of War," *Social Problems* 11 (Summer 1963): 75.

3. Judith Viorst, "Nuclear Threat Harms Children," *The Science News-Letter* 83 (February 16, 1963), 106.

4. M. Henriella Reinders, "Radiation Biology in the High School Course," 21 (February 1959): 60–61.

5. *Texans on the Alert* (Division of Defense and Disaster Relief, 1956), 18–19.

6. Donald J. Fluke, "Radiation and High School Teaching," *The American Biology Teacher* 22 (November 1960): 496.

7. *Nuclear Survival: A Resource Handbook*, University of the State of New York, State Education Department (1961), 37–38.

8. *Nuclear Survival: A Resource Handbook*, 39–40.

9. Robert L. Shrigley, "Fifth-Graders Explore the Atom," *The Elementary School Journal* 59 (February 1959): 277–281.

10. Milton Schwebel, "What Do They Think About War?" *Children and the Threat of Nuclear War* (New York: Meredith Press, 1964), 25.

11. Schwebel, "What Do They Think About War?" 29.

12. Schwebel, "What Do They Think About War?" 32.

13. Schwebel, "What Do They Think About War?" 25.

14. Tom Engelhardt, *The End of Victory Culture* (New York: Basic Books, 1995), 9.

15. Brock Chisholm, "Children, A.D. 2012," *Children and the Threat of Nuclear War* (New York: Meredith Press, 1964), 46–47.

Epilogue

1. "Children and Fear of War and Terrorism." *NASP Resources*, National Association of School Psychologists, 2002, www.nasponline.org.
2. *Terrorism: A War Without Borders* (Washington, D.C.: U.S. Department of State, 2002).
3. Erik W. Robelen, "Attacks, Causes, Aftermath Find Places in Some Lessons," *Education Week* 31, no. 2 (August 31, 2011): 14–16.
4. Jeremy Stoddard and Diana Hess, "9/11 and the War on Terror in Curricula and in State Standards Documents," Center for Information and Research on Civic Learning and Engagement, Tufts University, 2011.
5. *Learning from the Challenges of Our Times: Global Security, Terrorism and 9/11*, New Jersey Department of Education, 2003.
6. *National Strategy for Youth Preparedness Education*. Federal Emergency Management Administration and American Red Cross, 2014.
7. "Roots of Terrorism: Teachers Guide," *Public Broadcasting System*, 2003, www.pbs.org.
8. Ilene R. Berson and Michael J. Berson, "The Trauma of Terrorism: Helping Children Cope." *Social Education* (October 2001): 341–387.
9. "It Is Later Than We Think!" *Senior Scholastic*, 17 September 1945, 8.

Bibliography

Abraham, Herbert. "A World Organization for Peace." *School Life* 28 (October 1945): 3–4.

"Adjustment." *New Yorker* 18 August 1945: 17.

The Alert America Convoys. Valley Forge Foundation, 1952.

Amatora, S. Mary. "Emotional Stability of Children in the Atomic Age." *Education* 71, no. 7 (1951): 446–450.

"Another Citizenship Program." *The Journal of Education* 133 (May 1950): 135.

Ashby, Lyle. "Operation Crossroads." *Journal of the National Education Association* 35 (1946): 292.

"The Atomic Age." *Life* 20 August 1945, 32.

Atomic Energy and You. Los Angeles School District, 1953.

"Atomic Energy Book Exhibit for New York's Golden Jubilee." *Publishers Weekly*, 10 July 1948, 140.

"Atomic Energy Book Exhibit Touring the Country." *Publishers Weekly*, 27 November 1948, 204.

"Atomic Energy Here to Stay." *School Life* Supplement 31 (March 1949): 2.

Austin, Warren. "Atomic Weapons and World Peace." *Education* 71, no. 7 (1951): 414–419.

Benne, Kenneth. "Democratic Ethics in Social Engineering." *Progressive Education* 26, no. 7 (1949): 201–207.

Berson, Ilene R., and Michael J. Berson. "The Trauma of Terrorism: Helping Children Cope." *Social Education* (October 2001): 341–387.

"Birth of an Era." *Time* 13 August 1945: 17–18.

Blakely, R. J. "Living Without Fear in a Century of Continuing Crisis." *School Life* 35 (September 1953): 145–146.

Boyer, Paul. *By the Bomb's Early Light: American Thought and Culture at the Dawn of the Atomic Age*. New York: Pantheon, 1985.

Broder, Dorothy Elizabeth. "Life Adjustment Education: An Historical Study of a Program of the United States Office of Education, 1945–1954." Dissertation, Teachers College, Columbia University, 1976.

Brooks, John. "Ends and Means in Teaching a World Order." *Progressive Education* 30, no. 3 (1953): 72–74.

Brown, JoAnne. "A Is for Atom, B Is for Bomb." *The Journal of American History* 75 (June 1988): 68–90.

Burnett, R. Will. "The Teacher and Atomic Energy." *Education* 68 (1948): 545.

Burnett, R. Will, Ryland Crary, and Hubert Evans. "The Minds of Men." *School Life* 31 (March 1949): 11–13.

Burnett, R. Will, and Harold Hand. "Educational Implications of the Atomic Age." *Education* 71, no. 7 (1951): 429–445.

Bush, Merril. "World Organization or Atomic Destruction?" *School and Society* 23 (November 1946): 353–355.

Capen, Samuel. "The Truth Will Prevail." *School and Society* 9 (September 1947): 178.

Carey, Michael. "The Schools and Civil Defense: The Fifties Revisited." *Teachers College Record* 84, no. 1 (1982), 15–127.

Carr, William. "On the Waging of Peace." *Journal of the National Education Association* 36 (1947): 495–500.

Cartright, Sally. "When the Atom Bomb Was Born." *Progressive Education* 24 (October 1946): 4–6+.

"Children and Fear of War and Terrorism." *NASP Resources*. National Association of School Psychologists, 2002. www.nasponline.org.

Chisholm, Brock. "Children, A.D. 2012." *Children and the Threat of Nuclear War*. New York: Meredith Press, 1964: 46–57.

Christ, Henry. "The Atom Bomb Shakes the Classroom." *Journal of the National Education Association* 35 (1946): 296.

Civil Defense and the Schools. New York State Civil Defense Commission, 1953.

Civil Defense Educational Practices and References for Homemaking Classes, Classroom Practices. Washington, D.C.: Government Printing Office, 1957.

Civil Defense for Schools. (Pennsylvania) State Council of Civil Defense, 1952.

Civil Defense in Schools. Washington, D.C.: Federal Civil Defense Administration, 1951.

Civil Defense in Oregon Schools. Oregon State Civil Defense Agency, 1959.

Civil Defense Manual, Georgia Schools. Georgia Civil Defense Division, September 1952.

Clymer, Kenton. "The Ground Observer Corps: Public Relations and the Cold War in the 1950s." *Journal of Cold War Studies* 15, no. 1 (Winter 2013): 34–52.

Conant, James Bryant. *Education and Liberty: The Role of the Schools in a Modern Democracy*. Cambridge: Harvard University Press, 1953.

Corbally, John E., Jr. "The National Citizens Commission for Public Schools." *Educational Research Bulletin* 12 September 1956: 141–146.

Cousins, Norman. *Modern Man Is Obsolete*. New York: Viking, 1945.

Crary, Ryland. "Curriculum Adaptation to Changing Needs." *School Life* 35 (September 1953): 157–160.

Cremin, Lawrence. *American Education: The Metropolitan Experience, 1876–1980*. New York: Harper & Row, 1988.

Cremin, Lawrence. *Transformation of the School: Progressivism in American Education, 1896–1957*. New York: Alfred A. Knopf, 1961.

Dale, Edgar Dale. "It's a Nice World—Wasn't It?" *The High School Journal* 29 (October 1946): 176–180.

Davis, Ron. "Laboratory Practice in Protective Skills," *School Life* 35 (1953): 158–159.

Davis, Tracy C. *Stages of Emergency: Cold War Nuclear Civil Defense*. Durham: Duke University Press, 2007.

Day, Edmund. "Educational Mobilization in a Free Society." *The Educational Forum* 11 (November 1946): 10.

Diettert, Chester. "To Keep Democracy at Its Best." *School Activities* 21 (March 1950): 213.

Douglass, Harl. *Education for Life Adjustment: Its Meaning and Implication*. New York: Ronald Press, 1950.

Drake, Elizabeth, and Lillian Carmen. "A Broadcast for Brotherhood." *School Activities* 21 (October 1949): 56–58, 68.

"Duck and Cover: The Children Look at Atom Raid Drills." *Children and the Threat of Nuclear War*. New York: The Child Study Association of America, 1964: 34–45.

"Editorial." *Progressive Education* 24, no. 1 (1946): 16.

"Education and National Security." *Journal of the National Education Association* 41 (January 1952): 21–22.

Education for National Survival. Washington, D.C.: Government Printing Office, 1957.

Ehlers, Henry, ed. *Crucial Issues in Education: An Anthology*. New York: Holt, 1955.

Engelhardt, Tom. *The End of Victory Culture*. New York: Basic Books, 1995.

Escalona, Sibylle. "Children and the Threat of Nuclear War." *Children and the Threat of Nuclear War*. New York: The Child Study Association of America, 1964: 3–24.

Evans, Hubert, and Ryland Crary. "Atomic Education: A Continuing Challenge." *Teachers College Record* 50 (1949): 515–520

Evans, Hubert, Ryland Crary, and Glenn Hass. "Operation Atomic Vision." *Journal of the National Education Association* 37 (October 1948): 439–442.

Federal Civil Defense Administration. *Civil Defense Educational Practices and References for Homemaking Classes, Classroom Practices*. Washington, D.C.: Government Printing Office, 1957.

Fine, Benjamin. "'A Better World' Courses in New York City Schools." *The Education Digest* 12 (September 1946): 10–11.

Fleege, Urban. "The Teacher and Civil Defense." *Journal of the National Education Association* 40 (November 1951), 542–543.

Fluke, Donald J. "Radiation and High School Teaching." *The American Biology Teacher* 22 (November 1960): 496–497.

Flynn, George Q. *The Draft, 1940–1973*. Lawrence: University Press of Kansas, 1993.

Gable, Martha. "Philadelphia Classroom Television." *The Journal of Education* 134 (February 1951): 50–52.

Gail, Harry. "Atomic Energy and Education." *Progressive Education* 24, no. 4 (1947): 116–199+.

Gail, Harry. "Some Educational Implications of Atomic Energy." *Education* 67 (1947): 463–472.

Gallup, George. *The Gallup Poll: Public Opinion 1935–1971*. New York: Random House, 1972.

Garrison, Dee. *Bracing for Armageddon: Why Civil Defense Never Worked*. New York: Oxford University Press, 2006.

Gilbond, Florence. "The Impact of the Atomic Bomb on Education." *The Social Studies* 65 (March 1974): 109–114.

Gillette, B. Frank. "Atomic Energy: Resource Materials for Many Reasons." *The Clearing House* 25, no. 4 (December 1950): 205–207.

Glasheen, George. "What Schools Are Doing in Atomic Energy Education." *School Life* 35 (September 1953): 152–154+.

Goff, Aaron. "The Atom and Civilization: Ten Urgent Classroom Duties for Teachers." *The Clearing House* 21, no. 8 (April 1947): 457–460.

Gold, Milton J. "26 High Schools Use Radioisotopes: From Workshop Activity to Radioactivity." *The Clearing House* 28, no. 6 (February 1954): 359–362.

Gordon, C. Wayne. *The Social System of the High School*. New York: Free Press of Glencoe, 1957.

Goslin, Willard. "A Task for Administrators." *School Life* Supplement 31 (March 1949): 1.

Goulden, Joseph C. *The Best Years 1945–1950*. New York: Antheneum, 1976.

Gowing, Margaret. *Britain and Atomic Energy 1939–1945*. New York: St. Martin's, 1964.

Graebner, William. *The Age of Doubt: American Thought and Culture in the 1940s*. Boston: Twyane, 1991.

Graebner, William. *Coming of Age in Buffalo: Youth and Authority in the Postwar Era*. Philadelphia: Temple University Press, 1990.

Graebner, William. "The 'Containment' of Juvenile Delinquency: Social Engineering and American Youth Culture in the Postwar Era." *American Studies* 27 (Spring 1986): 81–97.

Graybeal, Lyman. "Democratic Education in a Time of Crisis." *School Activities* 20 (1949): 211–212, 215, 218.

Grossman, Andrew D. *Neither Dead Nor Red: Civilian Defense and American Political Development During the Early Cold War*. New York: Routledge, 2001.

Hales, Peter. "The Atomic Sublime." *American Studies* 32 (Spring 1991): 5–31.

Hand, Harold C. "Educating for Survival." *Educational Leadership* 4 (October 1946): 4–11.

Hand, Harold C. "Living in the Atomic Age: A Resource Unit for Teachers in Secondary School." *University of Illinois Bulletin* 23 (December 3, 1946).

Hardin Stearns, Virginia. "Denver's International Relations Club." *The Journal of Education* 133 (May 1950): 138–140.

Harmon, Millard. "Teaching Science in the Elementary School." *The Elementary School Journal* 50, no. 5 (January 1950): 273–276.

Hartmann, Andrew. *Education and the Cold War: The Battle for the American School*. New York: Palgrave McMillan, 2008.

Heffernan, Helen. "The School Curriculum in American Education." In Edgar Fuller and Jim Pearson, eds., *Education in the States: Nationwide Development Since 1900*. Washington, D.C.: National Education Association, 1969: 215–285.

Heil, Louis, and Joe Musial. "'Splitting the Atom'—Starring Dagwood and Blondie: How It Developed." *The Journal of Educational Sociology* 22 (January 1949): 301–336.

Hersey, John. *Hiroshima*. New York: Alfred K. Knopf, 1946.

Higbee, Homer. "The Social, Economic, and Political Implications of Atomic Energy." *Education* 71, no. 7 (1951): 420–428.

Hightower, Howard. "On War and Peace." *Progressive Education* 29, no. 7 (1952): 253.

Hilton, David, and Mary Jeffries. "Atomic Energy in the Classroom and Community." *Journal of Education* 131 (March 1948): 88–89.

Hitchcock, Richard. "Westinghouse Theater of Atoms." *The Journal of Educational Sociology* 22 (January 1949): 353–355.

Holliday, A.M. "The Atom and Educator." *Peabody Journal of Education* 25, no. 2 (September 1947): 102–110.

Hopkins, George. "Bombing and the American Conscience During World War II." *Historian* 28 (May 1966): 451–473.

"How a Small High School Meets the Challenge of the Atomic Age: Suffern High School Atomic Energy Club." *School Life* Supplement 35 (September 1953): 147+.

Hullfish H. Gordon. "Education in an Age of Anxiety." In H. Gordon Hullfish, ed., *Educational Freedom in an Age of Anxiety: Yearbook of the John Dewey Society 1953*. New York: Harper & Brothers, 1953: 208.

Hutchins, Robert. "The Issues in Education: 1946." *The Educational Record* 27 (1946): 365–375.

Interim Civil Defense Instructions for Schools and Colleges. Washington, D.C.: Government Printing Office, 1951.

"It Is Later Than We Think!" *Senior Scholastic*, 17 September 1945, 8.

Jacobs, Robert A. "Atomic Kids: Duck and Cover and Atomic Alert Teach American Children How to Survive Atomic Attack." *Film & History* 40, no. 1 (Spring 2010): 25–44.

Jacobs, Robert A. *The Dragon's Tail: Americans Face the Atomic Age*. Amherst: University of Massachusetts Press, 2010.

"Jam Session." *Senior Scholastic* 29 September 1948: 35.

Jenkins, David. "Social Engineering in Educational Change: An Outline of Method." *Progressive Education* 26, no. 7 (1949): 193–197.

Johnson, J. Clyde. "Teaching Democratic Skills and Attitudes." *The High School Journal* 35 (February 1952): 137–142.

Johnson, Jack T. "Protective Citizenship—Its Educational Implications." *School Life* 35 (September 1953): 150–151.

Kahler, Erich. "The Reality of Utopia." *The American Scholar* 15 (1946): 167–179.

Kane, William E. "An Atomic Age Week." *The School Review* 56, no. 5 (May 1948): 294–298.

Kaplan, Louis. "The Need for Creative Education." *The Journal of Education* 131 (November 1948): 241–242.

Kazdan, C. S. "Postwar Problems in Education." *Journal of Educational Psychology* 19, no. 6 (February 1946): 356–357.

Keesecker, Ward. "Duty of Teachers to Promote Ideals and Principles of American Democracy." *School Life* 30 (February 1948): 31–33.

Kennedy, Philip. "Oak Ridge and the Educational Crossroads." *National Association of Secondary-School Principals Bulletin* 30 (October 1946): 81–83.

Key, Lillian Wald. "Public Opinion and the Atom." *The Journal of Educational Sociology* 22 (January 1949): 356–362.

Kiernan, Denise. *The Girls of Atomic City*. New York: Simon & Schuster, 2013.

Kirshen, Rosalie. "A Unit on Atomic Energy in the Experience Curriculum." *High Points* 33 (February 1951): 27–31.

Lantagne, Joseph. "Health Interests of 10,000 Secondary School Students." *The Research Quarterly* 23 (October 1952): 330–346.

Laxson, Mary, and Berenice Mallory. "Education for Homemaking in Today's High School." *School Life* 32 (June 1950): 134–135, 138.

Learning from the Challenges of Our Times: Global Security, Terrorism and 9/11. New Jersey Department of Education, 2003.

Lilienthal, David. "Atomic Energy … and You." *Senior Scholastic*, 12 April 1948, 3.

Lilienthal, David. "Democracy and the Atom." *Progressive Education* 25, no. 3 (1948): 2–5+.

Lilienthal, David. "Education's Responsibilities." *School Life* Supplement 31 (March 1949): 1–2.

Lindsey, A. H. "I Taught Atomic Energy: With Statements by Thirteen Members of the Class." *Education* 71 (1951): 451–469.

Little, Stuart. "The Friendship Train: Citizenship and Postwar Culture, 1946–1949." *American Studies* 34 (Spring 1993): 35–67.

"Local High School Radio Forum." *School Activities* 22 (December 1950): 137.

Lora, Ronald. "Education: Schools as Crucible in Cold War America." In Robert Bremner and Gary Reichard, eds., *Reshaping America*. Columbus: Ohio State University, 1982: 237.

MacDougall, Curtis. "Language and Human Welfare." *Progressive Education* 25, no. 1 (1947): 269.

McClendon, Jonathan C., and Sylvia Davis Robinson. "Current Affairs for Social Studies Classes, 1956–1957." *The High School Journal* 40, 1 (October 1956): 38–44.

McClure, Dorothy. "Social-Studies Textbooks and Atomic Energy." *The School Review* 57 (December 1949): 540–546.

McClure, Dorothy, and Philip Johnson. "Where the School Takes Hold." *School Life* Supplement 31 (March 1949): 7–9+.

McEnaney, Laura. *Civil Defense Begins at Home: Militarization Meets Everyday Life in the Fifties*. Princeton, N.J., Princeton University Press, 2000.

McFarland, W.H. "World Unity in the Classroom." *The Journal of Education* 129 (March 1946): 96–97.

McGrath, Earl James. *Education: The Wellspring of Democracy*. Tuscaloosa: University of Alabama Press, 1951.

McMahon, Clara P. "Civil Defense and Education Goals." *The Elementary School Journal* 53 (April 1953): 440–442.

"Make and Show." *School Life* Supplement 31 (March 1949): 6.

Marshall, Kendric. "Teachers and the International Crisis." *School Life* 30 (June 1948): 2–3.

Mauth, L.J. "Prevention of Panic in Elementary-School Children." *The Journal of Education* 137, no. 2 (November 1954): 10–14.

May, Alonzo. "Atomic Energy and the Liberal Arts." *School and Society* 24 (August 1946): 131–133.

Meredith, Clyde W. "Civil Defense and the Schools." *School Life* 34 (April 1952), 99–100.

Metcalf, Harold H. "How Teach the United Nations." *The Phi Delta Kappan* 32, no. 4 (December 1950): 130–133.

Miner, Edwin. "National Conference on Zeal for American Democracy." *School Life* 30 (May 1948): 3–5.

"More Scripts for High School Radio Workshops." *Senior Scholastic*, Teacher Edition 26 Sept. 1951: 27T.

"Mouse Traps for Chain Reaction." *School Life* 32 (November 1949): 21–22.

Mygatt, Tracy. "World Government Is Common Sense." *Progressive Education* 24 (October 1946): 10–11.

National Strategy for Youth Preparedness Education. Federal Emergency Management Administration and American Red Cross, 2014.

Northcutt, Susan Stoudinger. "Women and the Bomb: Domestication of the Atomic Bomb in the United States." *International Social Science Review* 74, no. 3/4 (1999): 129–139.

Noyes, Richard. "The Teacher and the Atom Bomb." *Journal of the National Education Association* 35, no. 6 (1946): 296–297.

Nuclear Survival: A Resource Handbook. New York: University of the State of New York, State Education Department, 1961.

Ogden, W. "Ridge Kids Use the Atom." *New York Times Magazine*, 2 June 1946, 24–25.

"One World and the Teaching of History." *School and Society* 23 (August 1947): 132.

One World or None. New York: McGraw Hill, 1946.

Overstreet, Bonaro. "Understanding Our Fears." *Journal of the National Education Association* 41 (February 1952): 85–86.

Peckham, Earl K. "The Place of Civil Defense in Education." *School and Society*, 9 August 1952: 87–90.

Peffer, Nathaniel. "Politics Is Peace." *The American Scholar* 15 (1946): 160–166.

Perkins, John S. "Where is the Social Sciences' Atomic Bomb?" *School and Society* 17 (November 1945): 315–317.

Perry, Arnold. "Fundamental Education and the Defense of Democracy." *The High School Journal* 38 (January 1955): 117–123.

Peters, Charles. *Teaching High School History and Social Studies for Citizenship Training.* Coral Gables: University of Miami Press, 1948.

Peterson, Val. "Panic, the Ultimate Weapon?" *Collier's* 21 August 1953: 99–105.

Pike, Sumner. "The Promise of Atomic Energy." *Education* 71, no. 7 (1951): 407–413.

Pinette, Mattie. "School and Community Face the Atomic Age." *School Life* 35 (September 1953): 155.

"Psychologists Advise on the Atomic-Bomb Peril." *School and Society* 8 (June 1946): 405–406.

"Public Schools Urged to Stress Citizenship and United Nations." *The Journal of Education* 133 (November 1950): 239.

Rayburn, Sam. "That Civilization May Survive." *School Life* 28 (October 1945): 9–10.

Reaves, William. "Organized Extra-Curricular Activities in the High School." *The High School Journal* 34 (May 1951): 130–133.

Reinders, M. "Radiation Biology in the High School Course." *The American Biology Teacher* 21 (February 1959), 60–61.

Ridenour, Louis. "Science and Secrecy." *The American Scholar* 15 (1946): 147–153.

Robelen, Erik W. "Attacks, Causes, Aftermath Find Places in Some Lessons." *Education Week* 31, no. 2 (August 31, 2011): 14–16.

Robin, Richard. "Power and the Atom." *The Journal of Educational Sociology* 22 (January 1949): 350–352.

Roblee, Dana B. "What Schools Are Doing About Civil Defense." *School Life* 35 (September 1953): 152–159.

"Roots of Terrorism: Teachers Guide." *Public Broadcasting System*, 2003. www.pbs.org.

Rose, Kenneth D. *One Nation Underground: The Fallout Shelter in American Culture.* New York: New York University Press, 2001.

Rose, Peter. "The Public and the Threat of War." *Social Problems* 11 (Summer 1963): 75.

Rudy, Willis. *Schools in the Age of Mass Culture.* New York: Pantheon, 1969.

Rugg, Harold. "Progressive Education—Which Way?" *Progressive Education* 25, no. 4 (1948): 35–37, 45.

The Rural Civil Defense Youth Program. Office of Civil Defense Mobilization, circa 1959.

Russo, Anthony J. "A Unit Outline for Teachers: Civil Defense Instruction in Providence." *The Clearing House* 27, no. 9 (May 1953): 544–55.

Sawyer, Jeanette. "Science at Oak Ridge High School." *The Clearing House* 21, no. 6 (February 1947): 361–362.

Scheibach, Michael. *Atomic Narratives and American Youth: Coming of Age with the Atom, 1945–1955.* Jefferson, NC: McFarland, 2003.

Schwebel, Milton. "What Do They Think About War?" *Children and the Threat of Nuclear War* New York: Meredith Press, 1964: 25–33.

Schwebel, Milton, ed. *Behavior Science and Human Survival.* Palo Alto: Science and Behavior Books, 1965.

Sears, Laurence. "Anxiety in the United States of America." In H. Gordon Hullfish, ed., *Educational Freedom in an Age of Anxiety: Yearbook of the John Dewey Society 1953.* New York: Harper & Brothers, 1953.

Seyfert, W.C. "Youth in the Atomic Age." *School Review* 54 (June 1946): 319–320.

Shrigley, Robert L. "Fifth-Graders Explore the Atom." *The Elementary School Journal* 59 (February 1959): 277–281.

Spring, Joel. *The Sorting Machine: National Educational Policy Since 1945.* New York: David McKay, 1976.

Starie, John. "Schools and the Atom." *Education* 66 (1946): 501–502.

Starr, Benjamin, and Abraham Leavitt. "Social Studies and 'Operation Atomic Vision.'" *High Points* 31 (April 1949): 22–32.

State of New Jersey Division of Civil Defense. *Civil Defense and the School Principal.* (Trenton, Division of Civil Defense, 1952

Stearns, Virginia Hardin. "Denver's International Relations Club." *The Journal of Education* 133 (May 1950): 138–140.

Stoddard, Jeremy, and Diana Hess. "9/11 and the War on Terror in Curricula and in State Standards Documents," Center for Information and Research on Civic

Learning and Engagement, Tufts University, 2011.
Strong, Ruth. "Students Interpret the School to the Community." *The High School Journal* 34 (March 1951): 66–68.
Studebaker, John W. "Communism's Challenge to American Education." *School Life* 30 (February 1948): 1–7.
Studebaker, John W. *Education and the Fate of Democracy*. Los Angeles: University of California Press, 1948.
Studebaker, John W. "The High Schools of the Future." *School Life* 29 (April 1947): 306.
Studebaker, John W. "Secondary Education for A New World." *School Life* 29 (October 1946): 3–8.
Studebaker, Mabel. "The Teacher and the Atom." *School Life* Supplement 31 (March 1949): 1.
Survival in a Nuclear Attack: Plan for Protection from Radioactive Fallout. New York: New York State Civil Defense Commission, 1960.
Swan, Bryan, and Generose Dunn. "Unit on Atomic Energy for Junior High School." *The School Review* 62 (April 1954): 231–236.
Swomley, John, Jr. "Is the Military Invading the Boy Scouts?" *Progressive Education* 26, no. 3 (1949): 93–95.
"Teaching International Understanding." *The Phi Delta Kappan* 28, no. 2 (October 1946): 91–95.
Terrorism: A War Without Borders. Washington, D.C.: U.S. Department of State, 2002.
Texans on the Alert. Division of Defense and Disaster Relief, 1956, 18–19.
Todd, Lewis. "Atomic Energy and the Coming Revolution in Education." *School and Society* 62 (1945): 251–257.
Torrens, Hazel. "Current Events in the Ninth Grade." *The Education Digest* 12 (December 1946): 22–23.
"'Two to Five Years.'" *The Journal of Education* 128 (1945): 297.
Umstattd, J. G. "Contributions of the Secondary Schools in the Present World Situation." *The High School Journal* 34 (May 1951): 145–152.
Unruh, Adolph. "Life Adjustment Education—A Definition." *Progressive Education* 29, no. 4 (1952): 137–141.
Van Der Grinten, Willem J. "Not Only Science Teachers." *The Clearing House* 24, no. 8 (April 1950): 487.
Viorst, Judith. "Nuclear Threat Harms Children." *The Science News-Letter* 83 (February 16, 1963): 106–107.
Watson, Fletcher G. "Workshop on Atomic Energy: New England Project of Secondary Schools." *The Clearing House* 23, no. 8 (April 1949): 456–567.
Weitz, Leo. "A Social Studies Unit on Atomic Energy." *High Points* 31 (February 1949): 14–26.
Welles Sumner. "One Practical Chance—the UNO." *The American Scholar* 15 (1946): 141–143.
"What Is Operation Atomic Vision?" *National Association of Secondary-School Principals Bulletin* 32 (April 1948): 198–204.
White, Ralph. "Ultimate and New: Ultimate Democratic Values." *Progressive Education* 27, no. 6 (1950): 165–171.
Wiles, Kimball, and Woodrow Sugg. "Factors Influencing Curriculum Development." *Review of Educational Research* 24, no. 3 (June 1954): 195–203.
Winkler, Allan M. *Life Under a Cloud*. New York: Oxford University Press, 1993.
Wright, Quincy. "Barriers to World Peace." *School Review* 54 (1946): 576–583.
Yavanditti, Michael. "The American People and the Use of the Atomic Bomb on Japan: The 1940s." *Historian* 36 (February 1974): 224–226.
York, Barbara. "Quincy High School's P.D. Course." *The Journal of Education* 131 (September 1948): 218–219.
Zarlengo, Kristina. "Civilian Threat, the Suburban Citadel, and Atomic Age Women." *Signs* 24 (Summer 1999), 925–958.
Zeran, Franklin, ed. *Life Adjustment Education in Action*. New York: Chartwell House, 1953.
Zilliacus, Laurin. "World-Wide Union of Educators." *Progressive Education* 24, no. 2 (1946): 54–57.
Zwart, Gerrit. "How a Small High School Meets the Challenge of the Atomic Age." *School Life* Supplement 35 (September 1953): 147+.

Index

ABC's of Radiation 59
Abraham, Herbert 22
Alert America 14, 62–65
Allen, James E., Jr. 13
Amatora, S. Mary 87
The American Academy of Arts and Sciences 47
American Association of School Administrators 30
American Baruch Plan 83
American Education Federation 30, 33, 117, 122; *see also* Progressive Education Association
American Education Week 26
American Heritage Foundation 60
American Psychological Association 82
American Red Cross 155
Anacostia High School (Washington, D.C.) 56
Ashby, Lyle 19
Atomic-Age Combat 3
Atomic Attack 3
Atomic Energy Act of 1946 83
Atomic Energy Book Exhibit 60
Atomic Energy Club 56
Atomic Energy Commission 11, 13, 28, 47, 77, 83, 144
The Atomic Man 3
Atomic Spy 3
Atomic War 3
Atoms for Peace 82
Attack of the Crab Monsters 3
Austin, Warren 124

Baruch, Bernard 83
Baruch Proposal 83
Belmont (Ill.) High School 24
Benne, Kenneth 92
Benton, William 20
Berlin Blockade (1948–1949) 32
Berlin Crisis (1961) 148

Berlin Wall 148
Berson, Ilene R. 156
Berson, Michael J. 156
Bestor, Arthur 117
Bethe, Hans 16
Bikini Atoll 19, 51, 59
The Blob 3
Bloom Township High School (Chicago Heights, Ill.) 53
Boyer, Paul 10, 32, 75
Bradley, David, 82, 85
Brodie, Bernard 61
Brookhaven National Laboratory 58–59
Brooks, John 33
Brown, JoAnne 7
building drills 2
Burnett, R. Will 29, 31, 33–34, 124
Bush, Merril 22

Caldwell, Millard 64
Campaign for Peace 28
Campaign for World Government 19
Campus Elementary School (Ames, Iowa) 56
Capen, Samuel 119
Captain Video 3
Carr, William 28
Cathedral High School (Superior, Wis.) 144
Center for Information and Research on Civil Learning and Engagement (CIRCLE) 153
Central Intelligence Agency 92
China 1, 32
Chisholm, Brock 149
Christ, Henry 24
Churchill, Winston 1, 32, 119
Citizenship Education Project 121
Civil Defense Education Project 130
Cleveland Junior High School 46
Compton, Arthur 16

Index

CONELRAD 65
Constitution for a World Organization of the Teaching Profession 17
Corbett, Tom 3
Cousins, Norman 10, 61, 85
Crary, Ryland 31, 84, 126
Cremin, Lawrence 117
Cuban Missile Crisis 2, 7, 148

Dagwood Splits the Atom 60
Dale, Edgar 11, 13
Davis, Ron 121
Day, Edmund 26
Dewey, John 15
Diettert, Chester 117
dog tags 7
duck and cover drills 2, 8, 13, 67
Duke University Summer Institute in Radiation Biology 146

East High School (Denver, Colo.) 32
Einstein, Albert 13–16
Eisenhower, Dwight D. 3, 82, 130
Engelhardt, Tom 149
Eniwetok Atoll 2
Enola Gay 151
Escalona, Sibylle 9
Evans, Hubert 31, 83–84
Executive Order 10186 12

Families of September 11 154
Fat Man 1
Federal Civil Defense Act of 1950 12
Federal Civil Defense Administration (FCDA) 3, 7, 76, 78, 152; Civil Defense Education Project 12, 85; Educational Institutions Division 76
Federal Communications Commission 65
Federal Emergency Management Agency (FEMA) 155
flash drills 143
Flash Gordon 3, 45
Fleege, Urban 76
Fluke, Donald J. 146
The Fly 3
Forbidden Planet 3
Fordham University 115
Fort Hamilton High School (New York, N.Y.) 24
4 Action Initiative 154
Freedom Train 60

Gail, Henry 29–30
Georgia Civil Defense Division 88
Glasgow, Joe 50
Glenn, John 4

Goff, Aaron 8, 46–47
Golden Jubilee Exposition (New York) 60
Goslin, Willard 30–31
Gowing, Margaret 11
Graebner, William 13, 75
Graybeal, Lyman 32
Grossman, Andrew 75
Ground Observer Corps 61–62, 121
Gunsmoke 3

Hales, Peter 75
Hand, Harold C. 20–21, 33–34, 51
Harley School (Rochester, N.Y.) 50, 52
Harmon, Millard 54
Hartman, Andrew 15
Hass, C. Glen 83
Have Gun Will Travel 4
Hersey, John 10, 48, 61, 83, 85
Hess, Diana 154
Higbee, Homer 125
Hightower, Harold 128
Hiroshima 10, 48, 51, 75, 85
Hiroshima, Japan 1, 8, 32, 45, 49, 61, 84, 125, 151
Hiss, Alger 1
Hitler, Adolf 21, 119
Holladay, A. M. 10
Holmes, Robert D. 90
home economics 8, 14, 66, 76, 78, 80, 111, 146, 147
Hopkins, George 74
Hullfish, Gordon 79
Hutchins, Robert 19, 83
hydrogen bomb 1

The Incredible Shrinking Man 3
International Relations Clubs 32
Iron Curtain 119

Jacobs, Robert A. 1
James Monroe High School (New York, N.Y.) 84
Jenkins, David 92
Johnson, J. Clyde 127

Kahler, Erich 23
Kane, William E. 52
Kaplan, Louis 129
Kazdan, C.S. 11
Keene (New Hampshire) High School 58
Kennedy, John F. 2, 14, 147–148
Kennedy, Philip 49
Khrushchev, Nikita 2, 148
Korea 54
KPFA Radio Station 7

Laboratory School (Chicago, Illinois) 59
Liberty Science Center 154
Lilienthal, David 11, 13, 31, 77, 122
Little Boy 1, 16
Long, Daniel 61
Looking for Answers 155
Lora, Ronald 29
Los Angeles City School Division 56
Lynd, Albert 118

MacDougall, Curtis 119
Mad magazine 3
Making Atomic Energy a Blessing 59
"Man and the Atom" Exhibit 60
Manhattan Project 13, 16, 43, 47, 83
Mao Zedong 1, 32
Marshall, Kendric 122
Mauth, L. J. 87
May, Alonzo 24
McCarthy, Joe 55, 78, 87
McCarthyism 55, 78, 87
McFarland, W.H. 17, 121
McGrath, Earl James 128–129
McMahon, Clara 12, 79
Metcalf, Harold N. 53
Mount Baker High School (Deming, Washington) 56
mutually assured destruction 4
Mygatt, Tracy 19

Nagasaki, Japan 1, 8, 16, 45, 49, 51, 84, 125, 151
National Association of School Psychologists 151
National Association of Secondary-School Principals 26, 83
National Congress of Parents and Teachers 119
National Council for the Social Studies 120, 121, 153
National Council of Chief State School Officers 120
National Education Association 17, 28, 31, 80, 127; Research Division 80
National Parent-Teacher Association 77
National Plan for Civil Defense and Defense Mobilization 130
National Science Foundation 144
Neuman, Alfred E. 3
New England School Science Council 47
New Frontier 14
The New Jersey Commission on Holocaust Education 154
New York Committee on Atomic Information 60
New York State Civil Defense Commission 65, 74, 78, 92

New York State Education Department 47, 146
North Atlantic Treaty Organization (NATO) 1
Not of This Earth 3
Noyes, Richard 115

Oak Ridge, Tennessee 48; high school 48–49; youth council 48, 50
Office of Education of the Commission on Life-Adjustment Education for Youth 117
One World or None 16, 48
Operation Alert 3
Operation Atomic Vision 81, 83–84, 126
Operation Civic 28–29
Operation Classroom 28–29
Operation Crossroads 59
Operation Greenhouse 59
Operation Skywatch 61; *see also* Ground Observer Corps
Operation Teamwork 28–29
Oppenheimer, J. R. 16, 83
Oregon State Civil Defense Agency 90–91
Overstreet, Bonaro 77
Ozzie and Harriett 3

Peckham, Earl 66
Peffer, Nathaniel 21
Perkins, John 25–26
Perry, Arnold 128
Peters, Charles 121
Peterson, Val 78
Pike, Sumner 124
Platt, Dr. Joseph 52
Potsdam Conference 1
Proclamation on National Emergency 88
Proctor High School (Utica, N.Y.) 65
Progressive Education Association 30; *see also* American Education Federation
Prospect Heights High School (Brooklyn, N.Y.) 47
Public Broadcasting System 155–156

Quincy (Massachusetts) High School 121

Rayburn, Sam 19
Reaves, William 60
The Rebel 4
Reinders, Sister M. Henriella 144
Ridenour, Louis 23
Roosevelt, Franklin 74
Roots of Terrorism 155
Rudy, Willis 117
Rugg, Harold 30
Russo, Anthony 67

Saudi Time Bomb? 155
The Scholastic Book Service 82
Schwebel, Milton 149
Sears, Laurence 78
Seitz, Frederick 16
Senior Scholastic 80, 156
September 11, 2001 151–152
Shepherd, Alan 4
Smith, Mortimer 118
Society for the Advancement of Education 25
Society for the Psychological Study of Social Issues 82
Southern California Science Teachers Association 115
Soviet Union 1, 32, 130, 149
Sputnik 3
Stalin, Joseph 1
Starie, John 24
State Council of Civil Defense in Pennsylvania 45, 88
Stoddard, Jeremy 154
Studebaker, John W. 13–14, 26–28, 119, 121, 128
Studebaker, Mabel 31
Suffern (New York) High School 56
Superman 3
Szilard, Leo 16

Target America 155
Target Nevada 59
Telstar 4
Terrorism: A War Without Borders 152
Texas Division of Defense and Disaster Relief 144
Thompson, Elsa Knight 7
Todd, Lewis 16
Truman, Harry S. 1, 8, 12, 45, 115

U-235/U-238 16, 49
Umstattd, J. G. 123
United Nations 15, 16, 21–22, 39–40, 42, 53–54, 124, 135

United Nations Education, Scientific and Cultural Organization (UNESCO) 11, 39, 40
U.S. Children's Bureau 149
U.S. Office of Education 12, 80, 82, 120, 126
University High School (Springfield, Ill.) 55
University of Chicago's Laboratory School 59
Unlocking the Atom 59
Unruh, Adolph 117
Urey, Harold 16

Valley Forge Foundation 62
Van Der Grinten, Willem J. 8
Vietnam War 150
Viorst, Judith 144
V-J Day 16, 45

War of the Worlds 3
Weitz, Leo 47
Westminster College 32
White, Ralph 92
William Cullen Bryant High School (New York, N.Y.) 85
Williams, Grant 3
Wilson, Dr. Lewis 123
Winkler, Allan 74
WLNA Radio 56
World Conference of the Teaching Profession 17–18, 29, 35–44
World War III 3
Wright, Quincy 25

Yavenditti, Michael 74
Young Americans Day 65
Youth Council of the Atomic Crisis 48
youth forums 67

Zeal for American Democracy Program 120, 121
Ziferstein, Dr. Isidore 143